The Gentleman's New Jockey:
or,
Farrier's Approved Guide

The Gentleman's New Jockey:
or,
Farrier's Approved Guide

Edited with an Introduction by
Haley Ruffner

WHITLOCK PUBLISHING
Alfred, New York

The Gentleman's New Jockey: or, Farrier's Approved Guide first published 1700.

First Whitlock Publishing edition 2017

Whitlock Publishing
Alfred, New York
http://www.whitlockpublishing.com

Editorial matter @ Haley Ruffner

ISBN: 978-1-943115-24-2

This book was set in Adobe Garamond Pro on 55# acid-free paper that meets ANSI standards for archival quality.

Contents

Introduction i

Chronology of the Horse Industry x

Bibliography xii

Note on the Text and Acknowledgements xiii

The Gentleman's New Jockey:
or, Farrier's Approved Guide 1

Suggested Reading 174

Introduction

The Gentleman's New Jockey, a compilation of the late seventeenth century's newest animal husbandry techniques and remedies suggests using a broken sword blade to scrape sweat from one's horse, provides directions for inserting lead into a horse's forehead to create a white star, and examines the zodiac signs and lucky dates on which to administer important medical or breeding procedures. From horse-racing to breeding sheep, healing ointments to recipes for poison that blinds mice, this work offers a comprehensive guide to maintaining various livestock animals. This edition covers the horse industry, training, breeding, medicinal practices; killing vermin; and management of cattle, sheep, and swine. Published in London in 1700, *The Gentleman's New Jockey* acts as a manual for the eighteenth-century farmer and horseman.

Veterinary History

Mesopotamians first distinguished veterinarians from "witch doctors" and medical doctors in the 1800 B.C. Code of Hammurabi, which offered guidelines on the vocation. In this culture, veterinary medicine was a risky profession because "although every animal cured was worth a fee of a sixth of a silver shekel, if one should die under the ministrations of the Mesopotamian veterinary, he was required to pay over a quarter of the beast's value." Other ancient civilizations appear to have had some skill in treating equine health conditions. Horsemen known for their ferocity on the steppes between the Danube and Don rivers, the Scythians were able to geld stallions, cauterize wounds, and fit leather to horses' feet to prevent their going soft in the marshes.

Introduction

Greeks, though lacking the Scythians' skill, coined the term "hippiatra" for a specialist in equine medicine and organized their knowledge of Scythian veterinary techniques for further analysis. In the fifth century, Grecian works *De diaeta* and *De Prisca medicina* became the first known references connecting diet to health, a trend that has since exploded into today's fixation on healthy eating for both our livestock and ourselves[1]. The ancient Grecians successfully diagnosed various intestinal conditions, renal issues, typhoid fever, hernia, and tetanus, although they were less adept at treating these problems once identified. The Roman Empire contributed to this knowledge at first in the slaughterhouse, where they mandated inspection of meat to allay the spread of contagious diseases. This interest in equine anatomy expanded into further study of surgical operations, improving on preexisting methods as well as testing new ones. Horses were integral to the expansion and control of the empire, and their importance was reflected in the Romans' extensive assimilation of Greek and Spanish approaches[2].

Middle Ages veterinary medicine "sank to witchcraft" in Europe, although other parts of the world maintained dedicated animal care facilities, as in Indian equine hospitals, Hindu dedication to care of forest creatures, and meticulous Islam care for horses in observance of the Koran. In the fourteenth century, a farrier manual published in Padua, Italy became the basis for a collection of rules for treating horses and dogs. *Anatomy of the Horse*, penned by Carlo Ruini and illustrated by Leonardo da Vinci, was published in 1598, becoming one of the first volumes to detail the horse's body with some accuracy (Gianoli 66). Credited to Arnald of Villanova, the eleventh century poem *Salernitan Regime of Health* marked the revitalization of Western medicine. It was associated with the Salerno medical school in Southern Italy and existed in two hundred forty different versions and several languages, gaining widespread popularity because its basic theories were easy to memorize[3].

As scientific advancements progressed into medicine during the late century, the London Veterinary College was founded, marking

---

1 Curth 102
2 Gianoli 65
3 Curth 103

the beginning of a transition from treatments based on superstition to more beneficial medical practices. Some scholars argue that, before standard veterinary practices were established, "the most fortunate sick animals…were those left untreated"[4]. "Cures" for various ailments often included lead, mercury, bloodletting, and other substances or practices that are now understood to be carcinogenic or otherwise detrimental to a horse's health. However, in some cases, the remedies used at the turn of the century were effective; some are still in use today.

One important factor in determining equine health was the four humors (blood, yellow bile, black bile, and phlegm), the imbalance of which caused different maladies to afflict a horse. Purging excessive humors by way of bloodletting, sweating, forced vomiting, sneezing, emptying of the bowels, and enemas (then called "clysters") were considered necessary to maintain a horse's health, and are often cited as cures to various ailments in *The Gentleman's New Jockey*. Many veterinarians also recommended phlebotomy as a preventative measure, although the frequency and amount of blood proved controversial among practitioners. Another method of encouraging balanced humors included feeding or otherwise introducing substances into a horse's body, believed to boost a deficient humor or suppress a dominant one[5].

Despite the misguidedness of most treatments, the idea that the system of preventative medicine for horses during the eighteenth century was barbaric is also not true. Though mostly based on fallacy, medications to preserve and restore equine health were necessary to keep workhorses, racehorses, and pleasure horses in usable condition. A horse's worth was, and still is, primarily based on its ability to do a particular job, and preserving a well-trained, useful horse is more economical than replacing one lost to disease or injury. Today, when horses tread a fine line between pets and livestock animals, veterinary practices are often applied as a moral obligation out of sympathy to increase its comfort, even if the treatment will not increase a horse's value. This is a fairly recent development, the historical trend in medicine tending more towards financial efficiency than soft-heartedness. Rudimentary though they were, Early Modern medical manuals, including

---

4 From Farriery to Veterinary 44
5 Curth 124

INTRODUCTION

The Gentleman's New Jockey, did describe different states of sickness and health, catalogue diseases, and arrange them according to their supposed origin as in this work.

Classification of animals harks back to Aristotle's intellectual hierarchy, which "began with humans at the peak and animals and plants at various levels below according to their reasoning abilities"[6]. Domesticated animals—hardy, social, and easy to breed—fell above their wild counterparts on this scale, the latter being only suitable for hunting. Bestiary books indexing animals and their health were a popular genre beginning in the sixteenth century with the multi-volume work *History of Animals* by Swiss physician and naturalist Conrad Gesner, later translated by Edward Topsell into *The Historie of the Four-Footed Beaste*. These works promoted the study of animals based on their importance and productivity to "many arts, sciences, and occupations"[7].

Despite the suggestion that animals could be split into two groups—domesticated and wild, edible and non-edible, one valuable and the other useless—further examination suggests that this is an over-simplified lens with which to look at the relationship between man and beast. Books such as *The Gentleman's New Jockey*, which details such frivolities as creating markings on a plain-looking horse, indicates that livestock animals had value beyond their capability to work and provide for man; they had an aesthetic value as well. Inspecting the array of working animals included in the term "cattle" allows us to further understand the ways in which domesticated animals contributed to society. "Greater" cattle comprised horses, oxen, and cows, whereas "lesser" cattle encompassed sheep, swine, goats, and various poultry as defined in Topsell's translation. Within this definition, cattle can be further split into producers (of milk, eggs, wool, leather, and meat) and laborers (for pulling carts, plowing, travel, and riding for sport or in war)[8]. As agriculture was the most lucrative business in England, farm animals were indispensable for most people.

Although variations in England's geography meant that all regions were not suitable for keeping livestock, most farms had chickens, pigs,

---

6 Curth 16
7 Topsell 141
8 Curth 22

INTRODUCTION

and at least one horse. The horse's value as a work animal and for sport rendered its keeping worthwhile for nearly anyone. The majority of the population's ownership of such a complex animal merited the publication of health and training manuals focused almost exclusively on horses. Requiring more maintenance than a cow or smaller farm animal, horses warranted the higher cost and health upkeep due to the diversity of their use. As they moved through history from work animals to racehorses and pleasure mounts, and on to become a companion animal, horses became increasingly valuable as their worth expanded from utility to include a growing appreciation of their athleticism and beauty.

Publication of works such as *The Gentleman's New Jockey* also marked the growing literacy rate during the seventeenth and eighteenth centuries. Previously, farriers would have had a working knowledge of a few concoctions to administer during their travels trimming and shoeing horses, but standardization and knowledge within the general horse-owning public would be limited to word of mouth. In 1600, approximately thirty percent of men were literate, and only five to ten percent of women could read. By 1700, the male literacy rate had increased to nearly fifty percent, while women, largely barred from receiving a formal education, remained behind at twenty-five percent[9]. Even so, when *The Gentleman's New Jockey* was published, its audience would have been limited to mostly wealthy, well-educated men who owned horses.

A comprehensive education and licensing for veterinary practitioners, although considered essential today, was almost unheard of during the early modern period when *The Gentleman's New Jockey* was written. Very few "professionals" were university-educated, and historical evidence suggests that the practice of calling oneself a veterinary practitioner or farrier with no credentials whatsoever was common. During the early modern period, a healer need not have any medical knowledge or experience to treat animals[10]. The authors of *The Gentleman's New Jockey*, identified only by their initials, do not divulge the level of their education (or lack thereof) with regards to veterinary medicine, insisting that their work contains "the exactest Rules and Methods for Breeding and Managing horses" along with "approved Re-

---

9   Mitch 344
10 Curth 55

INTRODUCTION

ceipts and Remedies." By today's standards, such potentially groundless knowledge would not be considered a valid, reliable source of information, but in the late seventeenth and early eighteenth centuries this was standard.

The availability of a work encompassing training, care, and breeding of horses and other livestock animals enabled owners to consult an established set of expert-tested "cures" and take a more proactive role in their animals' health. Although said cures lacked in scientific backing and often did more harm than good, the fact that there existed a standardized, well-circulated manual over three hundred years ago implies that the theoretical application of medicine is much the same as the present day; the organization, format, and thorough descriptions of *The Gentleman's New Jockey* mirrors the arrangement of today's veterinary handbooks.

Racing and Breeding

The Gentleman's New Jockey instructs readers on the proper training and maintenance of "running-horses," with sections detailing diet, exercise, and performance-enhancing poultices and dressings. It also includes a chapter on making an ordinary horse into an "excellent Racer," which fits with the historical context of decreasing exclusivity in the racing world and private owners' ability to raise their own successful horses, a privilege once restricted to the monarch and his or her royal studs.

Horse racing began in England over a thousand years ago at weekly auctions in which crowds would gather to watch jockeys race sale horses to a marker and back. This exhibit of athleticism is described in the 1074 *Description of the City of London* by William Fitzstephen, who mentions that horse owners brought these sale horses to Smithfield to race, which allowed buyers to bet among themselves and choose the best and fastest horse[11].

The Crusades, beginning in 1095, brought news of Arabian horses to European horsemen, and in 1211 the first two Arabian stallions were imported to England. King Richard the Lion-Hearted instated racing at the Epsom Downs for Arabian horses, of which he was a

---

11   Gianoli 217

Introduction

dedicated admirer, and offered the first prize for racing: a sum of forty gold pieces. Henry VIII further encouraged racing as a sport, and in 1512 the Chester Fair directors offered a wooden bell for the winner of the horse race, later upgrading to a silver bell. Awarding silver bells as prizes eventually gave rise to the tradition of giving prizes for the first, second, and third place horses. The Lord Mayor of Chester found the bell to be poorly constructed, so he ordered a second one, with which he was still unsatisfied. The third bell proved satisfactory, and the Lord Mayor, left with three bells of varying quality, decided to give them out to the three fastest horses of the race.

Henry VIII also took steps to develop a system of breeding horses in England to improve the quality of native horses. He issued several proclamations and orders dictating the minimum height for studs, requiring dukes and archbishops to raise at least seven stallions measuring fourteen hands high (one hand is equal to four inches). He stationed royal studs at breeding centers established across the country, expanding their number when he discovered that the studs at Eltham, Kent, and Hampton Court Palace were not enough to keep up with demand for new stock[12].

Upon ascending to the throne in 1558, Elizabeth I ordered Italian horseman Prospero d'Osma to report on the breeding centers at Tutbury and Malmesbury. His work suggests some of the techniques for breeding that are used today and emphasizes the importance of choosing a location hospitable to raising colts, including specifics on rainfall, types of herbage, available shade, and the necessity of a facility capable of keeping foals warm in the winter and cool in the summer months. James I's reign brought a flood of Sottish noblemen to court in 1603 and, as they favored horse racing as a pastime, its popularity increased further. By 1617, James I requested a fence erected around the quarter-mile circular track in Lincoln to keep the spectators out and make the race's progress more clearly visible. He often watched races from a platform near the track.

This progress crumbled following Cromwell's rise to power and the fall of Charles I's monarchy. Since Cromwell, a strict Puritan, considered horse racing a sinful frivolity, he dismantled the industry. An-

---

12   218

INTRODUCTION

other motivation for closing down racetracks arose from suspicion of Royalists meeting at horse races to plot against Cromwell[13]. He and his cohorts sold the royal studs and dismissed the stable hands, the profit from their use draining into the Commonwealth's coffers. The tradition of annual spring and autumn races at Newmarket, established in the second year of Charles I's reign, fell by the wayside when Cromwell banned horse races in England[14]. However, the racing industry was revived during the Restoration—James d'Arcy, appointed Master of the Horse by Charles II, quickly assembled a sufficient herd of breeding stock to make up for Cromwell's dismantling the herd of prestigious royal studs[15]. Racing flourished during Charles II's rule; the king himself often rode as a jockey for sport.

Private owners of the previously-royal studs, having purchased them from Cromwell, began their own breeding farms and imported stallions from other countries[16]. Thus, the control of prestigious studs expanded from the royal family to include individual horsemen seeking to turn a profit and produce valuable colts. The effects of this shift are apparent in The Gentleman's New Jockey, which includes sections intended for private owners on the best ways to breed and raise a talented racehorse.

It was during William of Orange's reign that this edition of The Gentleman's New Jockey was penned, although the monarch himself contributed little to the horse industry, showing a marked disinterest in horse racing as compared to his predecessors and successor, Queen Anne. She was considered revolutionary as a woman who loved to hunt and showed a keen interest in sporting events. She bred and raced horses, instituting important races at her Ascot park in 1711, where her own stock often won. Seventeenth- and eighteenth-century monarchs of England set the precedent for the horse's value as an athlete in addition to its usefulness as a work animal.

The Gentleman's New Jockey filled a previously-sparse niche in the realm of animal husbandry, acting as a complete guide for several as-

---

13 Clee 54
14 Gianoli 221
15 219
16 221

Introduction

pects of farm life: breeding and raising livestock, caring for all manner of farm animals, training racehorses, and ridding one's house and barn of vermin. The remedies it prescribes for various illnesses and injuries are, although barbaric by today's standards and harmful in some cases, a contributing factor to modern veterinary manuals in how they detail and categorize different states of health. Its publication marks a shift in horse racing from an elite sport, reserved for royalty and nobles, to one accessible to any person with a working knowledge of horsemanship and a fast horse. *The Gentleman's New Jockey* was one of the first works of its kind to offer a thorough, concise volume of standardized livestock maintenance procedures to an increasingly literate public, thus developing into an influential manual for late seventeenth- and early eighteenth-century farmers throughout England.

Haley Ruffner, April 2017
Alfred, NY

Horse Industry Chronology

c. 4600 BCE	Horse bones and carvings interred in graves in Russia's Volga region
c. 3500 BCE	Horses domesticated in the Eurasian Steppes
c. 2000 BCE	Evidence of horses used for transport in chariot burials and warfare
1900 BCE	Papyrus of Kahun, first account of veterinary medicine, written in Egypt
1800 BCE	Code of Hammurabi distinguishes veterinarians from "witch doctors" and medical doctors
c. 200 CE	First horse race held in Yorkshire
c. 400 CE	Grecian works *De diaeta* and *De Prisca medicina* published: First known documents connecting diet to health
c. 475 CE	Farriers begin combining horseshoeing with doctoring
c. 800 CE	First written mention of "running-horses" in a record of a gift to King Athelstan of England
1211	First two Arabian horses imported to England
1356	Lord Mayor of London issues a request for all farriers within a 7-mile radius of London form a fellowship to regulate and improve practices
Early 1500s	Henry VII begins to import mares and studs and establish rules regulating horse breeding
1512	First record of a trophy given to the winner of a race
1519	Kiplingcotes Derby, the oldest horse race still in existence, established
1598	First comprehensive work on the anatomy of a horse, Carlo Ruini's *Anatomia del Cavallo, (Anatomy of the Horse),* published
1727	First written account of race results, John Cheney's *Historical list of all the Horse Matches run,*

	and all plates and prizes run for in England and Wales, published
1740	Parliament introduces an act to "restrain and prevent the excessive increase in horse racing," largely ignored by the racing industry
1750	Jockey Club established in Great Britain to create and enforce the Rules of Racing
1791	London Veterinary School founded

Bibliography

Clee, Nicholas. *Eclipse: The Horse that Changed Racing History Forever.* The Overlook Press, Peter Mayer Publishers, Inc., 2009. 15 Mar. 2017.

Curth, Louise Hill. *History of Science and Medicine Library: The Care of Brute Beasts: A Social and Cultural Study of Veterinary Medicine in Early Modern England.* Brill, 2009, ProQuest ebrary, site.ebrary.com/lib/herr/detail.action?docID=10419805. 8 Mar. 2017.

Gianoli, Luigi. *Horses and Horsemanship Through the Ages.* Crown Publishers, 1969.

Graham, Elspeth, et al. *The Horse as Cultural Icon: The Real and Symbolic Horse in the Early Modern World.* Brill, Vol. 18, Intersections, 2011, EBSCOhost, search.ebscohost.com/login.aspx?direct=true&db=e000xna&AN=399876&site=ehost-live. 10 Mar. 2017.

Wear, Andrew. *Knowledge and Practice in English Medicine, 1550–1680.* Cambridge University Press, 2000, ProQuest ebrary, site.ebrary.com/lib/herr/detail.action?docID=5004608. 10 Mar. 2017.

Acknowledgements

Thank you to Dr. Grove for tolerating (and answering) the barrage of questions it took to complete this book, for your eternal patience in providing formatting help, and for always encouraging my best work.

In addition, thank you to my family for your endless support in all my endeavours, and to Ellie Woznica for your companionship in the long hours spent in the library during this project.

Note on the Text

Based on the fourth edition of *The Gentleman's New Jockey*, this work reflects the grammatical conventions of the late seventeenth century. Archaic spellings and phrasings were preserved in order to maintain the authenticity of the literature, as have inconsistent spellings such as "juyce" and "juice," used interchangably throughout the text. Obvious typographical errors have been corrected.

The
Gentleman's New Jockey:
OR,
Farrier's Approved Guide:

CONTAINING THE EXACTEST RULES AND METHODS for Breeding and Managing HORSES, in order to bring them up in the best manner for Profit, Pleasure, Service or Recreation; especially in what relates to Racing or Running, Coursing, Travel and War; with Directions for Heats, Dieting, and Dressing, and the several Paces requisite on any of the like Occasions.

Also approved Receipts and Remedies for all manner of Diseases, Sorrances, Hurts, or Grievances incident to Horses, according to the Opinions of the best Farriers of all Ages. With Directions for preventing Sicknesses and Griefs, and the Symptoms whereby they are known.

To which is added, A Second PART;

Containing many rare and new Secrets, relating as well to Management and Cure, and what else may contribute to the Advantages of Buying or Selling; and many other Things and Matters, highly necessary to be known by all Lovers of good Horses.

Illustrated with sundry Curious and Necessary CUTS.

The Fourth Edition, with large additions.

Liscens'd and Enter'd, according to Order.

London: Printed by *W. Onley,* for *Nicholas Boddington,* at the Sign of the Golden ball, in *Duck lane,* 1700.

Advertisement

An useful Concordance to the Holy Bible, with the various Acceptations contain'd in the Scriptures, and Marks to distinguish Commands, Promises, and Threatnings. Also a curious Collection of Similes, Synonumous Phrases, and Prophecies, relating to the Call of the Jews, and the Glory that shall be in the Latter Days. Lastly, The Titles, Appellations given to Christ and the Church, not in any Concordance yet extant. Begun by the Industrious Labour of Mr. Vavasor Powel, *and Finish'd by Mr.* N. P. *and Mr.* J. F. *Recommended to the Studious Reader, by the Reverend* John Owen, *D. D. The Supplements being plac'd, in this Impression, in their proper Places. The Fourth Edition, with Additions.* London: *Printed for* Eleanor Smith, *and sold by* Nicholas Boddington, *at the* Golden-ball *in* Duck-lane.

The Preface

to the

Reader.

Reader,

Though Sundry Persons of no Small Experience, have undertaken to give Mankind an Insight into the Manner and Method of bringing Horses to the height of Perfection, in what relates to their Rule and Management on those Occasions of Services to which they are properly intended or designed; as likewise to know and distinguish them in all their Capacities, with Directions for discovery and cure of Distempers, Grievances and Defects incident to them; yet, let me tell you, there are many things material of that kind that have not as yet been made Publick, all, or most of them, so necessary to be known by the curious Enquirers into this Mystery, that without knowing them no Man can be an excellent Jockey, or an expert Farrier, nor consequently to have his Judgment approved in any thing material, relating to Horses or Horsemanship. Great indeed have been the Studies and Travel of the Industrious of many Kingdoms and Provinces in this Affair, whose Labours have worthily found Acceptance: the Consideration of which has imboldned me to make this Essay, hoping it will turn to a general Advantage, being rightly considered, put in practice, and applied: since that by a more than usual Curiosity, what has been contained in this, has been searched out through all the revealed and (till now) unrealed Mysteries of this Kind, so that the Reader may with easie prospect or survey, find out whatsoever he can wish, relating to a good Horse, good Management, knowledge in the Craft, and means to preserve the Beast from a multitude of Infirmities; nay, rescue him from Death and utter Destruction, near and into which, through Neglect or Ignorance, he is fallen, and thereby not only give themselves Satisfaction in the prolonging the Days and Health of so noble a Creature for Service and Employment sundry ways, but always redound to their Profit and Advantage in capacitating them to sell and dispose of, at a considerable Rate, the Horse that they perhaps deemed lost and of no Value: In order to which, beside the Mystery of Breeding Horses to the best advantage, and regard that is to be observed in Diet, Dressing, and preparing them for Racing, and other the like Undertakings, with habit of Body, knowledge of Goodness or Badness; and in general and particular, all things of that or any other kind whatsoever that is requisite to be known, I have set down such ready and easie Ways and Methods to discover Diseases and other Grievances jointly and severally; as also speedily, and with little Charge, Labour, or Trouble to Prevent, Redress, or Cure them; that what is written most undoubtedly satisfie the largest Expectation as to things of this Nature, or at least give that Satisfaction, that the most Penurious or Critical may find enough to make him confess that time and cost were not employed in vain. Some indeed may upon the first sight object against the Smallness of this Book, and imagine a Subject so copious, allowing such Variety, is unlikely to be cramed in so narrow a Corner: But let me tell them, that when at great Expence and Pains they have searched larger Volumes, as also in the unpublished Practices of those who are famous at this Day, they will find that they have wearied themselves in vain, in Travelling the longest Way about, when in all that tedious Progress they found nothing material but what is comprehended in this Treatise. But not longer to Gloss upon what is able to speak for itself, I shall only say, that of this Kind, nothing exceeds it; I'll so submit it to the Censure of the ingenious Reader, and remain,

Yours further to oblige you,

G. L.

To his very good Friend, the Ingenious *G. L.* upon the Perusal of his Book, intituled, *The Gentleman's New Jockey,* &c.

Sir, I have view'd the Authors of this kind,
That did pretend or seem'd to this inclin'd; (This Art I would have said) such as have wrote,
Unskill'd themselves, and would have others taught;
But prove in all their large Pretence Defect,
Either the cause of Weakness or Neglect.
But in the Work you to the World propose,
I find to th' Essence of those numerous those,
Who vainly Built on a Foundation bad;
Which from ill Structures of old Times they had,
You have additions in strange measure made,
Such as those Ages never knew. So when
Light from the Chaos sprung, those things were seen,
That brooding Darkness spreads her Wings upon,
And long onscur'd beneath her gloomy Throne.
In Epitome, the noble Creature you
Have in Perfection given all his due;
From his beginning to his end, and done
What has in Ages past, attention won
By all the Brave, the Wise, when in a Cloud
It dimly shone; but you remove the Shroud,
And give, unscreen'd, a prospect of what may
Avail Mankind, and your own Worth display:
Whilst those who Hunt, who Travel, or in War,
At Home seek Fame, or in the Regions far,
Aided by your Advice, as ever free
From Danger great, or Ill-conveniency.
And could a Creature made Irrational,
Arrive to that Discretion which we call
The Sence of Mankind; he far more than they
Would for this Work, a humble Tribute pay,
And at your feet himself an Offring lay.

Sir, yours to serve you, J. D.

The Contents of the First Part

Chap. I.	Excellent Rules and Instructions to be observed in the Election and Choice of good Breeders. How and at what time a Mare ought to be Covered; how ordered being with Foal, and after Foaling: with many other things of the like nature. 1
Chap. II.	How to manage a good Colt to bring him to be an excellent Horse, and what is to be observed in the timely knowledge of what he will be when come to Maturity; with the season of Weaning, Cutting, and Usage, &c. 4
Chap. III.	When and how to Break a Colt and render him tame and gentle; with the Art of Management on that occasion; and other things requisite to be known. .6
Chap. IV.	The Art of Dressing in general, for bringing a Horse to Perfection, and render a Groom acceptable....7
Chap. V.	How to render any tolerable good Horse an excellent Racer; and how Running-horses ought to be managed in Diet, Physick, and Excercise. 8
Chap. VI.	A Continuation of what is to be done to the Running or Race-horse, relating as to his Election, Dressing, Feeding, and other Usage, &c......................10
Chap. VII.	Heats, what they are, and how to be managed to the best Advantage, in Racing..........................12
Chap VIII.	Bread of the first Make, and what ought to be observed in the Distribution thereof. 13
Chap. IX.	Other things to be taken notice of, in regard to the Running-horse, in the second Fortnight........15
Chap. X.	The Running-horse's third and forth Fortnight; and as to Diet, Usage, and bringing him to the Weighing-post. 17
Chap. XI.	What the Rider ought to be; Backing and good Management; the Office and Duty of a Groom. ...19
Chap XII.	Infallible Rules for Buying a good Horse; and how so well by sundry Marks and Tokens to distinguish Good from Bad, as well in case of soundness as Diseases. 21

The Contents

Chap. XIII. *Infallible Rules to know the Age of a Horse by his Teeth, in all their particular and general Marks; as also by the Tail.* 28

Chap. XIV. *How exactly to know the State and Condition of a Horse's Body relating to Fatness or Leanness, Health or Sickness.* 30

Chap. XV. *Observations to be taken as to the State of a Horse's body from the privy-parts, limbs, feeding and other matters.* 33

Chap. XVI. *Of the Elementary Parts of a Horse's Body, and of the Agreement of Humours therewith. A Discourse of Corruption and Generation, as to Goodness or Badness, Health or Sickness.* 34

Chap. XVII. *How a Horse ought to be used in general and particular, as to his Physick, Diet, and looking to.* ...37

Chap. XVIII. *An exact Description of the Veins of a Horse, how situate in the Body; as also of Blood letting; and how, and upon what Account of Sickness or other Defect, they are to be opened, for the prevention of Death or Danger.* 40

Chap. XIX. *A Discourse of the Sinews and their situation, with their use and office; and what in that kind is to be observed as to the state of a Horse's Body; with a Description of the Bones, &c.* 42

Chap. XX. *Of Blood-letting in general, and how to know when and where it is convenient to let blood, for preventing of Sickness, or recovery of Health.* 44

Chap. XXI. *A Description of Diseases, Grievances, or Sorrances incident to Horses; to foresee them by sundry Signs and Tokens, and know whence they arise; with the Ways and Methods of Preventing, Redressing and Curing them, by approved Rules and Remedies.* 46

Chap. XXII. *Excellent Receipts for the Cure of Diseases, &c. in Horses or Mares, according to the best Experience of Skillful Practitioners.* 48

The Glaunders, from what it proceeds, and how to cure it.

The Quinzy, what it is, and how to cure it.

The Contents

A Horse's Bleeding at the Nose, how to stay or prevent it.

For Pains in the Teeth or Jaws.

The Canker in the Nose, what it is, and the Remedy.

A Remedy for the Colick, Belly-ach, or Belly-binding.

The Lasks or Bloody-flux, and its Remedy.

The Botts, what they are, and their Cure.

For the Shoulder-strain, a good Remedy.

Broken-wind, what it is, and to remedy it, if not past cure.

A Horse burnt by a Mare, how to cure.

For a dry Cough, Cold, Pursiveness, Broken-wind or Shortness of Breath.

To restore decayed and putrified Lungs.

A dry Consumption, its Remedy.

The Breast pain, from whence it proceeds, and how to cure it.

Heart-sickness, or Antecor, *whence it proceeds, and how to remedy it.*

Foundring in the Body, or Surfeiting, how occasioned, together with the Remedy.

The Greedy-worm, or Hungry-evil, what it is, and how to be remedied.

Yellow and Black-jaundice in a Horse, the cause and cure.

Costiveness, from whence it proceeds, and its Remedy.

The Cramp, or Convulsion in the Nerves or Sinews, how occasioned, together with the Remedy.

The Mourning of the Chine, its cause and the means to cure it.

Frenzy or Madness, its cause, with the means to remedy it according to the experienced way.

The Falling-evil, its cause and cure.

The Sleepy-evil, what it is, and the way to remedy it.

The Horse-pestilence, and its cure.

For Chest-foundring, the Remedy.

The Contents

For an Obstruction in the Bladder, or Windiness in the Bowels, use this approved Medicine.

The Pole-evil, how to know and cure.

The Fistula, how to discover and cure.

Hard Kernels under the Throat, how to remove them.

For the Navel-gall, the Remedy.

For a Blow, Bruise, of the like Misfortune that causes a Swelling or Tumour, the Remedy.

The Scratches, their Remedy.

For Foot-foundring, a Remedy.

For the Canker in the Head, a Remedy.

For the Mange, or dry Scurvey, a Remedy.

The Vives, and their Remedy.

For Swanking in the Back, or a Strain in the Kidneys, caused by indiscret Riding, or Over-burthening.

For Pains or Foulness in the Reins or Kidneys, an excellent Scowring.

Relief for an Attaint or Over-reach on the Heel, or the like.

For the Water-falcion, a Remedy.

A Cure for the Sorrance called the Ring-bone.

For the Ives, a Remedy.

To take off the Film or Skin from a Horse's Eye.

A Mallender, the Remedy.

For the Palsie or Apoplexy, a Remedy.

A Farcion in the Head and Neck, to cure.

A Lineament to cleanse a Wound, new or old.

For Kibed-heels, commonly called the Mules, a Remedy.

The Quitter-bone, what it is, with its Remedy.

An excellent Cure for the Blood-spavin.

For a Neather, Attaint or Over-reach in the Pastern joynt.

For a Frush the cure.

To dissolve the Humours, and thereby anticipate Diseases.

The Contents

To soften any hard Swelling or contracted Hardness.

For the Splint, Wind-gall, or Bladders of Gelly, in or about any of the Joynts subject thereto.

To cleanse any putrified or other Sore, the best way.

A Horse Planet-struck, how to cure.

For the Poze, or excessive Cold, a Remedy.

To remedy Hideboundness.

For a wet inward Cough, a Remedy.

For a dangerous Cough, commonly called the Dry Cough.

For the Yard of a Horse falling, a Remedy.

To Prevent the Mattering of the Yard.

Of two Diseases incident mostly to Mares, and known by the name of the Pestilent-consumption.

How a Mare that is subject to cast her Foal ought to be used.

How to oblige a Mare to cast her Foal.

Of the several Fevers in a Horse, and how to cure them.

For any Disease or Stoppage in the Liver.

For a Botch or Sorrance in the Groin of a Horse, a cure.

For a general Manginess, the Remedy.

The Barbs, what they are, and how removed.

For Blood-shot-eyes, an excellent Remedy.

For any Film, Bite or Blow in the Eye, a Remedy.

To kill Lice, or remove Flies from offending your Horse.

To rid a Horse from any foulness or disorder in the Body.

For an extraordinary Bruise or Bite.

For a Horse that is troubled with the Strangling, a cure.

To remedy the Swelling of a Horse upon having eaten any infectious thing in his Provender, that may prove dangerous.

For a Heart-burning or Wasting occasioned thereby.

To prevent staling blood, a Remedy.

The Contents

Another Remedy for the Farcy, vulgerly called the Fashon.

An Approved Cure for any Sinew-strain or Over-reach.

Diseases in the Hoof how to remedy, and first for a Horse, upon sundry occasions is apt to cast his Hoof.

Hoof-bound; what it is, and how to remedy it.

To soften or harden the Hoof, the best way.

To harden a Hoof as occasion requires.

For any Hurt or cankerous Sore in or on the Hoof.

To oblige a Horse to carry his Ears well, the way.

For the Grievance called the Frounce, *a cure.*

For a Heat, which sometimes occasions a breaking out in the mouth and lips, the cure.

Woolf-teeth what they are, and how to ease their Pains.

To stanch any Bleeding, a speedy way.

To supply the defect of the falling of the Crest, a remedy.

To cure Manginess, or the like Disorder in the Crest.

The Navel-gall, what it is, and its Remedy.

For a Sitfast, or horney Excressence under the Saddle, whereby the Horse is disabled from carrying it in good order as he ought.

For any Knob or Wen near the Saddle-skirt, or the Sides of the Horse, a Remedy.

For Weakness in the Back, a strengthening Remedy.

For the swelling in the Cods of a Horse naturally, or occasioned by any Bruise.

For Burstenness, or the Rupture in a Horse.

The Art of discovering hidden Griefs or Ailments in a Horse, and from whence they proceed.

The Bone-spavin, what it is, &c.

For a Haw in the Eye.

For the falling of the Fundament, a Remedy.

To preserve a Hoof from decaying.

Interfering, what it is, and the Remedy.

A false Quarter, what it is.

The Contents

The Melt on the Heel.

An excellent Remedy for any Strain or Swelling.

How to remedy the harm done a Horse by unadvisedly and unskillfully letting Blood.

For the Leprosie in Horses a Remedy.

For any Disease in the Lungs, an excellent Remedy.

For the Swelling of the Horse's Legs, a cure.

The Flying-worm, what it is, and how to cure it.

Excellent Directions for the preventing of Diseases in Horses, sundry times, &c.

A Cure for a sore or defective Mouth.

For the Mellet in the Heels, a cure.

The Stavers, their signs, cause and cure.

For the Stone, a Remedy.

To take away a Wen in the Neck or any part of the Body.

The Crownet-scab, what it is, together with the cure.

To draw out a Thorn or Stump, or any Iron or sharp thing gotten into the Flesh.

Strain in the Coffin-joynt or Socket of the Hoof.

For a Horse that is perpetually sick and out of order, by retaining a spice of former Surfeits.

For a Cold in the Summer when the Horse is defective in filling, or by too rank feeding.

A Through-splint or Screw-pin, what it is, and the cure.

To render a brittle Hoof firm and serviceable.

To cure the Anbury.

To prevent a Horse's pissing Blood, an excellent Remedy.

For a broken Knee, the Remedy.

For the Dropsie.

To joyn a Sinew that is out.

For a Wound or Hurt in the Tongue, a Remedy.

For the Itch in the Tail, or any other part.

The Contents

Another excellent Remedy for a Tetter.

For a Hurt or Wring in the Withers.

Worms of any sort in the Body of a Horse, how to kill them.

A Salve for any sort of Wound, how to make and apply it.

An Imposthume, to ripen or cure.

For any Internal Sickness, another good Remedy, never before published.

For the Ach-weakness and Numbness of the Joynts.

How to stay any violent Luseness.

The Lampas, what it is, and how to cure it.

The Fig in the Foot of a Horse, the cure.

The Flanks, a disease so called, and the way to remedy it.

The Shackle-gall, and its cure.

A Horse that is gravelled, how to remedy.

Of Cataplasms or Poultises.

For any Swelling, Imposthume, Rankling, Wound or broken Bone, a Cataplasm or Poultise.

For a Swelling in the Throat or under the Ears.

To draw or break a Boil or Ulcerous Sore.

An excellent Poultise to asswage any Pain or Tumor.

A Poultise to break any infectious Sore.

To disperse the Flux or Oppression of Blood in any part.

Imposthume or sudden Swelling in any part of the Body.

In case of the Palsey in the head, (a Disease seldom happening in Horses) a Poultice.

Rare and New Experiments.

To heal and contract any Wound.

An inward Balm to destroy Worms, and heal internal Bruises.

An excellent Balm in case of any Sprain, Internal Bruise, Swelling, old Sore or Gun shot.

An excellent Balm to be given a Horse inward, in case

The Contents

of a Consumption.

An excellent Red Water to cure Ulcers.

An excellent Water to allay any Internal Heat or Feavourish dispondency.

In case a Horse be troubled with the Stone, an excellent Water.

For any Disease in the Eyes, another excellent Water, &c.

An excellent Purgation for Gravel in the Bladder or Kidneys.

For the Ulceration of the Yard, an excellent Water.

An excellent Powder for the Falling-sickness or Falling-evil in a Horse, &c.

A Powder for the Ague, which frequently happens, especially in young Horses.

To Purge Choler and Flegm, an excellent Remedy.

An excellent Bath to allay any Swelling, or such-like Disorder, especially Diseases in the Legs, occasioned by the descending of evil Humours.

A Bath to soften and mollifie the Skin.

To stanch Blood in any Vein or Artery.

An excellent Medicament to provoke a Horse to Vomit, as also to purge the Belly.

An excellent Purge, good on sundry occasions.

For a Joynt-sickness.

For an Internal Ulcer.

An Excellent Electuary for a dangerous Cough or ratling Cold.

Chap. XXIII.	*The Symptoms of Diseases, &c. in general and particular, how to foresee them, and prevent them, as also to know when they happen, &c.*...................109
Chap. XXIV.	*Directions for Making and Preparing Ointments, Oils, Salves, Waters, Purgations, Poultises, Charges, Supplements, Pills, Powders, &c. singular good in case of any Distemper or Sorrance; many of them never before made publick.*......................111

An Ointment to search any Wound or Ulcerated Sore,

The Contents

or any thing of the like nature.

An Ointment excellent good in case of Botches, Boils, Scabs, or the like.

To skin any Wound, an excellent Ointment.

To mollifie and asswage any Swelling, an excellent Ointment.

An excellent Ointment to cool and allay any Inflammation.

An excellent Ointment, or rather Balsam to be inwardly given a Horse for Obstructions, Bruises, and other Ailments and Grievances.

To make a green Ointment, proved by Experience more effectual than what has formerly been published.

An excellent Remedy for the Staggers, or any Pain that suddenly takes a Horse, sometimes to the loss of his Life.

An excellent Salve for any Wound.

An approved Salve, to draw any Stub, Thorn, or Splinter of Bone or Wood out of the Flesh.

To fill a Wound, Ulcer or the like with good Flesh, a Plaister.

A Salve to draw Corruption from the bottom of any Wound, or to draw a Swelling, or any such Grievance to a Head.

To ripen any Tumor or asswage any Swelling not abounding with extraordinary Humours.

An excellent Poultise for a Tumour or Swelling.

A Charge to ease a Pain in the Back, or for any Sprain.

To mollifie any Chap or rough Sore.

For the Eyes of a Horse afflicted by any means, an approved Water to cure or ease them.

A Water to wash the Mouth in case of any Sorrance or Defect.

A Pill good for any Internal Disorder.

A Supplement exceeding good for any Strain or Grief in the Sinews.

A Vomit for a Horse that has a queesie Stomach, thereby to render him a good appetite.

To purge Melancholy.

The Contents

To purge Phlegm.

Chap. XXV. *Clysters, how to make them, and on what occasion they ought to be applied in order to their effectual working and bringing away bad Humours.* *117*

In case of any Pestilential Disease, occasioned by cholerick or fiery Humours.

For any Internal Distemper proceeding from Melancholy.

For any Distemper Internal, occasioned by sanguine corrupt Blood, or watry Humours, by means of bad Concoction or Obstruction, &c.

For Sickness in general, an approved Clyster.

In case of Restringency or hard Binding.

Chap. XXVI. *Cordials, Cordial pills, Drinks, and Drenches, Purgations and Suppositories, wonderfully conducing to the health and strength of a Horse.* *119*

Diapente, *an excellent Powder in case of any Cold or Pestilential Diseases.*

An excellent Cordial ball to be given in case of any Internal Distemper, and especially to prevent the Consumption or wasting any part.

An excellent Drench to cure any Internal Distemper proceeding from any of the four Humours of the Body, &c. *especially such as are Pestilential.*

Suppositories, and their Use.

Chap. XXVII. *Perfumes, Baths and Purgations, what they are, to what end they serve; with the manner how, and under what considerations they ought to be applied* *121*

Directions for Purging, according to the estate and condition of your Horse.

Chap. XXVIII. *Costicks, Corrosives and Rowelling; what they are, for what cause, and in what manner to be applied.*.. *123*

Rowelling, what it is, and how to be performed.

The Contents of the Second Part

The manner of Breaking a Horse the best way, and Perfecting him in his Paces, &c. and preserving him from Danger, &c.

Chap. II.   How the Jockies make old Horses look young; a lean Horse artificially and naturally, how fatned by Jockies. A Remedy for Restiffness, Neighing, and the Vice of Lying down in the Water; the Art of making Stars, Snips, Blazes, setting on false Ears, Tails, Manes, &c. with a discovery of many other Secrets..................131

To make a Horse that is really old, seem young.

A Horse subject to lie down in the Water, how to remedy it.

A Tired or Restiff Horse, to remedy.

To prevent the troublesomeness of a Horse's Neighing, which may prove disadvantageous to the Master, especially in the time of War.

If a Horse be dull, and will not feel the Spur without much wounding, take the following Directions to make him go very nimble with or without a Spur.

To make a lean Horse artificially fat, or to seem so to the Buyer.

To make a lean Horse really fat, the best and cheapest way.

To make the Hair of a Horse, that stands rough and staring, smooth and sleek.

To make Hair come where it is thin, or take it away where it is thick.

Stars, Blazes, Snips, what they are, and how to make them for Ornament or Disguise in any part of the Horse where it may be conveniently situate. To make a black Star, Blaze, or Snip, in a white Horse.

To make a Blaze-Royal.

To loss of Ears, how to supply.

Chap. III.   How to set a Horse off for Sale to the best Advantage, by Trimming, Washing, &c. as also Directions for the Management of a Horse in Hunting, relating to his Leaping, &c. with other things and matters worthy of note..139

The Contents

How to make a Ball, wherewith a Horse being well lathered and smoothed down, shall look exceeding sleek and comely.

How to manage a Horse in Leaping, taking a Hedge, Gate, Stile or Ditch, &c.

Chap. IV. *What the Stable to keep a good Horse in ought to be, and how he ought to be regarded; the Hoofs how to be corrected and mended in Shooing, and upon other occasions. 142*

Critical Days, and the Observations thereon.

Some further Consideration upon the Cause of Diseases, and how to remove them, Physically discussed, &c.

The Spirits, what they are, with their office, &c.

A further Description of the External Parts, &c.

The Contents of the Third Part

How to know if a Beast be sound or not, as also to know if an Ox or a Cow be sound or whole of Body.

How to Fat an Ox.

An order or course how to Fat Oxen in the Stall.

How a Man may be rightly informed for to Buy and Sell oxen.

How to keep a Cow that is great-bellied with Calf.

Of a Cow that wants Milk, having but lately Calved.

A good way to Cut or Geld a Calf.

To Rear and for to Breed Calves for Increase.

To help the Garget in the Throat of a Beast.

To Cure the Garget on the Tongue.

Against the Garget coming by any Push.

Against the Garget in the Maw.

To Kill Lice and Ticks in Cattle.

For the Murrain in Cattle.

For the Flux or Lasks.

For the Lasks, or Ray in Calves, or cough in your Bullocks or Heifers.

For Scalds or Manginess.

For any Distemper in the Lungs.

For Pissing Blood.

For the Taint and Gargets.

For stoppage of Urine.

For any poisonous Infection, or pain in the Belly, swelling or internal Bruise.

For the Sperenges and Staggers.

For being Hide bound, which hinders the Growth of Cattle.

For the Feaver in Cattle.

To Cure Halting.

To Cure the Swelling in Cattle by breaking into fresh Pasture, and over-feeding, or by licking up some venomous Matter.

For any Blain, or outward Sorrance.

For the Pains in the Bowels.

For the Quinzy.

The Contents

Of Rams, Ewes and Lambs, &c. 163
To chuse good Breeders for the increase of Sheep.
The best time for Covering.
How, after casting, to order your Lambs.
For the Scab or Mange.
For bruised Joynts, broken Claws, &c.
For the Rot or Plague.
For any Disease in the Lungs of Sheep.
For the Head-ach, or pains in the Head of Sheep.
In case of Rheums, Catarrhs, or Coughs.
For the Plague and Rot, another.
For Boils, Aposthumes, or Ulcers, that are not come to a head.
For Scabs, or Breaking out of that nature.
For Pursiveness.
For St. Anthony's Fire.
To supple broken Joynts, Sprains, Wrenches and fractured Bones.
For Lameness, which is occasioned by too much Wool growing in the fleshy part of their Feet.

Of Swine, their Breeding, Ordering, &c. 166
How to have a good Breed of Swine.
To Cure Pains in the Head or Teeth.
For the Head-ach, or Sleepy-evil.
An Approved Remedy for the Measles.
For the Swine-Pox.
For Rheums, or Catarrhs.
For the Plague, or any Disease in the Melt.
To Cure the Flux.
For the Belly-ach.
For the Diseases in the Eyes of Swine.

For Killing of Vermine... 169
To Kill Rats and Mice.
To make Rats and Mice blind.

The Contents

To drive away Rats and Mice from a House or other place.

To gather together all the Rats and Mice into one place in a House or Barn, and to kill them.

To catch Moles.

To Kill Moles.

To Kill Weasles.

To prevent Weasles from the sucking of Eggs.

To drive the Snakes and Adders out of the Garden.

To Kill Snakes and Adders.

To Kill Pismires.

To Kill Bugs.

To Kill Fleas.

To Kill Lice.

To Kill Crab-lice.

For Nits and Lice in the Head.

To Kill Caterpillers.

To Kill Flies.

To gather the Flies together.

To keep Cattle from Injuries by Flies.

The Art of taking Fish.

To take all sorts of Birds, Fowls, &c.

The Gentleman's New Jockey:
or,
Farrier's Approved Guide

Part I

Chapter I

Excellent Rules and Instructions to be observed in the Election and Choice of good Breeders. How and at what time, &c. a Mare ought to be covered; how ordered being with Foal, and after Foaling: With many other things of the like nature.

When any Gentleman or other Person is desirous of a good Breed of Horses or Mares, that may redound to his Pleasure and Profit, the chief thing is to Elect a Stallion and Mare or Mares, in whom Nature has not been defective, but rather fitted and framed for Beauty, and promising Service of what kind soever, shall be premeditated or proposed by the Master: And in this case Instructions, how unerringly to chuse such as may answer the largest Expectation, will not be amiss; for as the Sire and Dam are, such will the Colts prove.

Those that have been Curious in these Matters, generally agree, That the Mare for Breed ought to be about four Years old, altogether free from Diseases and Sorrances, clear-limbed and well-proportioned, her Eyes lively and sparkling, standing somewhat out of her Head, her Ears standing direct, her Buttocks broad and well fleshed, with a large Womb, little Head, well-set Shoulders, and an arched Neck; her Legs even, and her Neighing sharp and clear, with fair Hoofs and large Ribs, one that is not used to trip, stumble, interfere, or given to any untoward Vice. And, to answer her, the Stallion ought to have a little Head, inclining to leanness, a swelling Forehead, full Ears, with Eyes quick and standing out, being somewhat speckled with Blood, his Nose crook'd, or bending, his Nostrils wide, and Neck arch-wise, with a large Throple and curling Mane, broad and well-set Shoulder, large Knee-joynts and lean Legs, a well-set Chest, broad Back, large Ribs, a gaunt Belly, Straits, Fillets, and good Pasterns well knit and well proportioned; of Colour bright Bay or Cole black; tho' in this case other Colours may prove good, both as to the Mare and Stallion, as will appear when I come to describe the Goodness by the Colour, &c.

A good choice thus made, it will be altogether convenient to feed your Stallion for a time in the Stable with heartning Provender, as Splent-beans, Oats, or sod Barley, to render him lustly and mettlesome; yet let him feed on good green Pasture, at least a Day before you suffer him to back your Mare, for fear that by reason of the Stiffness of his Limbs, and pursiveness, he break his Wind, by overstraining himself in the Action. If the Mare, who must be suffered to Run in good warm Pasture, appear averse, and by her striking, or otherwise seeming unwilling, decline the company of the Stallion, then will it be convenient to put some little Stone Nag to wooe her, taking away the Stallion; and when he has brought her to a compliance, take him away without suffering him to leap her, and bring the Stallion to her again about Sunrise, the Ground being well fenced about, and no doubt, she will suffer herself to be Covered; and if the Horse be very headstrong and unruly, it will not be amiss if two strong Men lead him to her, and manage him according to their Discretion, for fear he injure himself or the Mare: Then let a third stand ready with a Bucket of Water to cast upon the Mare's Shape with all his force, so soon as the Horse dismounts her, thereby to make her shrink up her Body, and suddenly close the Womb for the better retention of the Seed, and in fifteen Days after, if you imagine she has not thorowly conceived, which will appear by her turning her Back part to the Wind, pricking up her Ears often, scenting the Air and Neighing, you may put the Horse to her again, and order her in the like manner; but if the Mare be notwithstanding defective in Conception, to remedy that fault, occasioned mostly by too much Blood, take a Pint of Blood from either side her Neck four or five Days before your Horse covers her; and the Day after bleeding, take a Quart of warm Milk, half a Pint of the Juyce of Mugwort, *London*-Treacle and sweet Butter, of each two Ounces, dissolve them together; and, being well mixed, give them to her in a Drenching-horn luke-warm, and you shall find the effects answer your expectation, especially if twice or thrice repeated the succeeding Mornings when she is fasting: Now, if she has conceived well, her Belly within four or five Days will appear gaunt, her Hair more bright and shining, her regard of Horses or Noises will be little, her Ears will flag, and she will decrease in Flesh. Ways there are to oblige either Stallion or Mare, in case of Averseness, by rubbing a Spunge in the Mare's Shape, and with it rubbing the Horse's Nose; as likewise by rubbing the shape of the Mare with *Aqua-vitæ*, Cow-itch, Nettle-seed, or the like. But these being vulgarly known, I shall wave them, and proceed to what is more material to be known.

Some are of opinion that knitting the right Stone of the Horse with a List or silken Cord, will occasion a Female Colt, and the contrary a Male; but how often that has failed, there are few Horse-keepers ignorant: Wherefore, to follow Experience, the best Mistress, observe, that if you would be sure of a Stone-Colt, to let your Mare be covered when *Aries, Taurus, Gemini, Cancer* or *Leo*, are predominant, which are called the Masculine Signs, and as well rule Irrational as Rational Creatures; and if your Desires are contrary, then let *Virgo, Libra, Scorpio, Sagittarius, Capricorn, Aquaries* and *Pisces* be your Rule: For as Heat and Moisture are the great distinguishers of the Sex, so the Signs influencing more or less, are not a little concerned as Second Causes in the production of Things participating of their Temperature.

For the space of three Weeks, or somewhat longer, after Covering, she must or ought to be kept in a warm House and dieted, lest the Conception happen to be impaired before the Colt be well formed; and after that, she must not be hard rid nor laboured, neither come into any Ground or Stable where unruly Cattle are, or where, by attempting to leap, she may injure herself, lest thereby her Master's Expectation be frustrated by Miscarriage, or the Colt's beig bruised and distorted in the Womb, observing the Ground she runs in be dry, well sheltered with spreading Trees, Barns, Quick-sets, or Out-houses, under or into which she may retire at leisure; nor will it be unnecessary to have Racks with Provender placed where she may advantageously come at them.

The time that a Mare goes with Fold, unless some mischance happen, is by the most curious Observers a Year and ten Days, though some will have it twelve Months, within eleven Days, and some a shorter time: This, I confess, as to a young mare, in her first Teaming, does often fall out; but upon the second Foaling, rarely or never: For if your Mare be covered about the middle of *December*, which is the best time, she may have the advantage of the Summer to run in, and the following Spring to bring up her Colt; she will be sure to Foal about the latter end of *December* in the following Year, as has been frequently observed. When the time of her Foaling is at hand, let those, whose charge it is, be assisting to her in case of Emergency, she being continued in a warm House without any Halter or other cumbersome thing about her, whereby she may be hindred, or caused to slip or stumble, being eased in her lying down with clean Litter, or laying, to receive the Colt, if the cast is standing, as often it happeneth; and when she has dried the Colt with licking, let the Milk be drawn from her Teats before it be suffered to suck her, to prevent its clotting in the Udder; and if she be scanty of Milk, to prevent driness and increase it, give her warm Mashes, with some Powder of Brimstone and Water, wherein Scabeous or Vervine have been boiled, anointing her Dugs with the Decoction of Lavender and Spike; and so a Month or two after you may work or rider her gently; but beware if she return hot, let her cool before the Colt take the Teat, lest the Humours being too much stirred, the heat of the Milk being too greedily taken, surfeit him: and thus a good Mare will bring forth able Colts till ten Years, and a Stallion get them till twelve; but those that on either part happen beyond, will prove Weaklings, and not worth the Rearing. And thus much for procuring a Colt; which, if carefully reared, will unquestionably make a good Horse or Mare.

Chapter II

How to Manage a good Colt to bring him to be an excellent Horse, and what is to be observed in the timely knowledge of what he will be when come to Maturity; with the Season of Weaning, Cutting and Usage, &c.

THE FIRST THING NECESSARY TO KNOW, whether a Colt will make a Horse, and when arrived to Maturity, is the Observation of his

Lineaments; the Measures in that kind to be taken thus: *Viz.* Observe whether the Shin-bones be strait and even, of a convenient thickness, according to a durable or firm proportion, that thereby his tallness and ability may be perspicuous or apparent; for in consideration of the proportion between the Withers and the Knee you may expect a duplicity, or as much more in the height, when at full growth: Observe likewise that his Joynts be large, that he be full boned, his Head small, lean, and his Ears of a moderate size, standing upright; and that when he Stands in due proportion or at ease, his hinder Legs stand rather backward, than crimpling forward, or even, which denotes speed; then consider by his striking or playing up and down, his shaking his Ears or Tail, staring in your Face, or the like, his mettlesomeness, and consequently thereby his other Properties, as far as his tender Age will give an insight; for in this Creature, as in all others of what kind soever, there remains no certainty without Industry, as to the reducing them to the Service of Mankind: and to prevent his pining when weaned, let him by all means be out of the hearing of his Dam; and that he may the better brook her Absence, make a mixture of the best Butter you can get, and Rhubarb, Rue, or Savine, and give him nourishing Diet; as Mashed, Splent-beans, boiled Oats and hot Bran, with somewhat that is green and juicey, for the space of three Months after: Nor indeed ought he to be used to hard Provender altogether, till three Years after his weaning, lest thereby he be induced to neglect his feeding, and so by pining hinder his growth and activity.

If it be a Stone-Colt, separate him from that of the other Kind at the expiration of a Year, lest he spoil himself in attempting what he is not capable of performing; and he better to prevent the endangering of either, is to put them into indifferent Pasture, well fenced, to prevent leaping, and secured by Quick-sets, trees, Barns or other Out-houses from heat and cold.

If you design to make Geldings of your Stone-Colts, the proper time to Cut or Dilapidate them is, when they are between nine and twelve Days old, if the Sign of Life, which may be gathered from the Course of the Moon in her Wain, being either in *Aries* or *Virgo,* especially in the Spring or Fall, which are accounted the fastest Seasons; tho', by those who well understand this Affair, it may be done at any time, the Weather being dry, and not too hot or cold.

And what now remains as to the bringing a Colt to perfection, is the Art of the Breeder, which may be, as the vulgar term has it, the principal Verb; for if he neglect his Charge, all that has been mentioned may prove ineffectual; which Breeder I also term the Manager; for it is not enough to feed a Horse well, that thereby he may be able-limbed and high-metled, but to bring him up to those degrees of Exercise, that may fit him for any Imployment.

Chapter III

When and how to break a Colt and render him tame and gentle; with the Art of Management on that occasion: And other things requisite to be known.

When your Colt comes to be of the Age of five Years, which is a good time to oblige him to endure Backing, or the like, having before made your self familiar with him, by stroaking, feeding, and shaking the Bridle at his Head, laying your Hand on his Back, smacking the Whip or cheriping, with such other Signs as are commonly used to create a sufferance; lay your Saddle, not being over-weighty, on his Back, continuing, as you observe his Patience, either to increase the weight by a greater Saddle, or by placing on the lesser somewhat that is ponderous, stroaking and rubbing him the mean while, that he may not be intent on what you are doing; and when you have brought him to a moderate sufferance, girt the Saddle with soft Girths, and that by degrees, loosning them as you find occasion, by his winching and untowardness, or straitning them according to his patience or sufferance: And so in case of your Bridle, which ought to be the second thing applicable in reducing a Colt or young Horse to obedience and observance of what tends to his Accomplishment; but if you perceive a Colt, before he arrive at any convenient maturity, to be over-mettlesome, it will be prudent to make him familiar, tho' at a Year old, yet not to press his Back with Weights, for fear of swaying it; and so by a timely taming him, render the Men of other Nations mistaken, who account the *Spanish, Turkey, Barbary, Pelopenesian, Flanders, Friesland, Holland* and *Artois* Horses and Mares the best; and the reason that they give is, That they may be made gentle at any Age; which the *English* Horses, *&c.* whom they term of a stubborn Nature, cannot be brought to: And indeed the taking up too soon, without due Management, is the occasion of stunting, and other ways spoling of many a brave Colt or Horse.

If you find your Endeavour in the manner aforesaid fail you in bringing a young Horse or Mare to your bent, then consider Loneliness may be the Occasion, or Fear more than Obstinacy: To remedy which, oblige him or her to accompany those that are tame, and of a gentle nature, already reduced to what you design them, or at least flexible on that occasion; and by often observing that you handle them, a mildness will be created, and a familiarity will increase to a desire of the like; nor will moderate rubbing and picking the Feet be least means to introduce what else you desire. Some indeed, and those the most curious in this Affair, will not attempt to Back a Horse which they intend for Travel or Racing till the sixth Year; but, in my Opinion, they lose a Year's Service thereby to no manner or purpose; for the indifference as to the one and the other in this kind, is inconsiderable, unless in case of extraordinary Labour; for consequently the longer the Horse remains in good Feeding undisturbed, the stronger he grows in Joynt and Member, tho' less in Swiftness.

Chapter IV

The Art of Dressing in general, for bringing a Horse to Perfection, and render a Groom acceptable.

As the old Proverb is, *The eye of the Master maketh the Horse fat*; so his Care in this case is necessary (if he cannot be present as he would wish) to chuse such a one, who may, as well in his absence as in his presence, perform his Duty to the benefit of his Master and his own credit, by rising early and tending the Horse or Horses within his Charge; and the best Diet in this case to bring him to perfection is, in general, a Quart of Splent-beans, and two Quarts of Oats sweet and well sifted; and so, if more than one, the like quantity; then Curry him well with an Iron-comb; after that, Dust him over with a Dusting-cloth, and pass a French Brush upon him, and smooth him with your Hands to sleekness, passing over him before a Woolen, and then Linnen-cloth, cleaning his Eyes, Ears, Sheath, and picking his Feet, placing his Mane decently, and smoothing his Forehead after its being ruffled, to loosen the Skin; when, putting on the Saddle, with Wisps under the Firths, ride him gently to Water, suffering him to dung or stale as you see him inclined; and in case of defect, provoke him thereto, by letting him scent Horse-dung, musty Hay, cast Straw, or entring into Stubble or Stable, which ought rather to be Paved than Planked, to prevent dampness and unwholesome scents arising from the Urine, *&c.* that stagnates between the crevices or chinks of the Plank, be they laid never so close; of which I shall say more, when I come to give Directions for a convenient Stable, fitting for choice Horses.

Your Horse being in the Stable, rub him down in all Points, as before directed, ever observing to begin at the Head, and so proceed backwards, till every part be decently passed over, not taking off the Cloaths, or removing the Bridle under the space of an Hour, and in the mean while let him have the same quantity of Provender as before, adding, as he has dispatched the Grain, a Bottle of sweet Hay, bound up hard and cast in to the Rack, that with some labour he may pull it thence. These are General Rules and no way to be contemned.

Chapter V

How to render any tolerable good Horse an excellent Racer; and how Running-Horses ought to be managed in Diet, Physick, and Exercise.

Seeing Racing is highly in esteem, and a good Horse of that kind much coveted by the Gentry of all Nations, I shall give an insight into that Affair: And first in chusing a Horse for Running; consider he be well managed, tractable, and no ways skittish, familiar with his Keeper or Masters, and free in eating what is given him; lively and sprightful in his Looks and Actions; let him be one of an indifferent large Reach, well set in the Shoulders, and fairly Hipped, gaunt and smooth Backed, his Legs and Thighs standing in due proportion, his Head small and lean, with fast or firm Cheek-bones, a sharp or Hawk's Nose, wide Nostrils, and wide Throple, and not exceeding twelve Years old, nor under six, being of a good breed: Nor is this all, for the main point of bringing him into a condition to Run successfully is yet behind, which consists in Dieting, Dressing, and carefully Ordering, before he undertake such a Business: and in that case, If you would have him answer your expectation, observe these Rules, which are by the great Searchers into these Mysteries, held to be the best and most Exact that ever were.

Resolving to Run, and having marched your Horse, the first thing you ought to do is, to have a regard to the state and condition wherein he is, as to what relates to his Body in case of health, sickness, or any other advantage or disadvantage, and especially to what follows: As, whether he be newly taken from Grass or Soil, and by that means his Body rendred foul and unwieldy, over-burthened and incumbred with unnecessary Fatness or crudy Humours; or, on the contrary, whether by bad usage he is become Poor and Feeble; or that some hidden Infirmity occasional Leanness or a Pining-away: And, lastly, whether he be in good liking or state for your purpose, by having been moderately exercised and carefully regarded.

These Considerations weighed, if you find your Horse in the first Estate, let him not be matched, if it will any way stand with your conveniency, under six Weeks, or at least not in less than a Month, that so he may be prepared by regular Diet and Applications, for what he is to undertake.

In the second Estate, unless your Horse be neglected, six Weeks or a Month may suffice, for that his Exercise may be suited to his Feeding, and in that case no time lost, if he that looks after him be careful and industrious.

As for the last Estate, it being a medium between what has been mentioned, a Month, with care is a sufficient Time to bring him to Perfection, tho' many allow six Weeks; but that is, as he is more or less inclining to Fatness or gross Humours, contracted in rank Pasture, or by eating foul Provender. Now if it happen, that any of these kinds be of a free and fiery Nature, apt to excess in Exercise, thereby to spend the Flesh he has got in Dieting, he that has the Charge, must, as he sees convenient, restrain him with a hard Hand; and if he be defective in feeding, let him use what devices he can to prompt him to it, as by whistling, singing, stroaking, clapping, rubbing, tossing his Provender, and the like, though in that fat Horse this ought chiefly to be observed; for the lean one, and that in good Case, if not indisposed by any distemper or grievance, will eat freely and may be better exercised, not needing any restriction, unless upon extraordinary occasions, though a tender regard must be had to either.

Chapter VI

A Continuation of what is to be done to the Running or Racehorse, relating as to his Election, Dressing, Feeding, and other usage, &c.

THE HORSE YOU CHOSE, though naturally a good one, having contracted fatness or foulness by the means of too rank a Pasture or bad Feeding; having put him in a dry and warm Stable, though somewhat darkish, to render him the better feeder, lead him out by the rising of the Sun, suffering your Boy or Servant to give him a turn or two till you have cleaned the Stable of what is offensive; then return him, and having your Bridle dipped in Beer, put it on, and tye him up to the Rack; at what time Curry him in every part, beginning with his Head, and ending with his Pasterns; then with a Horse-tail, or clean Dusting-cloth, dust him over; that done, take your French-brush and rub him decently over in all parts, by such degrees as he may be sensible of your kind Usage: As, first the Forehead, then the Cheek, so down the Neck on either side, after that the Shoulders, Fore-legs, Back, Sides, Belly, Buttock, Rump, Thighs, Cambrils and the very Fetlocks, beating and sounding the Hoof to make it fit the firmer: This done, pass over him again with your wet Hand to settle the Hair, and render him sleek and shining; the wet being dried by a repetition, cleanse his Sheath, Cods and Tuel, his Ears, Eyes and Nostrils, his Mane, Foretop and Tail, and rub the moisture, if any remain, from between his hinder Thighs, picking his Hoofs, and running over his Legs with dry Wisps, or a clean Woollen-cloth, which if you see convenient, pass over his Body, combing his Mane, Tail and Foretop: having thus far proceeded, take a large Body-cloth or Kersey, or the like thick Woollen, and spread over him, if it happen in the Winter; but if in the Summer, that of a lighter make will suffice; on that lay a light saddle, girting it pretty strait, yet let a Wisp or two stick on each side to give him the better breathing room, bracing the Cloth likewise about his breast and shoulders.

Thus having dressed and attired your Horse, spurt a little warm Beer into his Mouth, and lead him forth, and mount him, leaving one to order the Stable and provide good Litter against your return, Wheat-straw being the best of that with which tour Stable must always be sufficiently provided for; other Straw is not only unseemly, but unwholesome; the Oat-straw breeding dislike, and the Barley-straw, if your Horse be subject to eat it, a scowring.

Being abroad, ride your Horse for some time a foot-pace, which, by the Curious, is termed Racing; and after you have so managed him for the space of a Mile or more upon firm Ground, advance to a Hill that gently rises, if the situation of the place afford it; and mending by degrees his pace, bring him to Gallop up; the which when he has performed, Lead him, or Race him down to the best advantage, suffering him to cool and contract an equal temperature of Body. Thus having done, as you see your Horse without over-training will conveniently bear it, lead him a Mile or thereabouts to some pleasant River or Spring, and suffer him to Drink moderately, and then Exercise him as before; which done, give him a second Watering, and after that another Exercise, ever observing to Exercise him before and after Watering; which done, Ride him easily home, and coming to the Stable door alight, suffering your Horse to stale or dung in the foul Litter; the which if he refuse freely to do, provoke him to it by whistling, clapping, or waving your Switch, raising the Straw under him, and reaching him upon it; which will often, if not always, oblige him to it, custom giving him an insight into what you would have him to do. This done, lead him in and fix him upon the clean Litter, take off the Saddle and Body-cloth, rubbing him down, and cleansing him as before; after which put on the Cloths, girt him gently with Circingles; and for the easement of his lying down, put Wisps between of loose and soft Hay or Straw: being thus cloathed, pick his Feet, and stop them with Cow-dung, casting into the Rack a small bottle of Hay well tied up, It being well dusted, and let him tear it out at leisure, whilst he stands on his Bridle.

Your Horse having stood for the space of an Hour, rub his Head well with a Hempen cloth; and having cleansed the Manger draw his

Bridle, and take about three Pints of large white Oats well dressed, sweet and dry, free from light Oats, or such as are defective; for the preparing of which, you ought to have a small Wire-seive: These being given to your Horse, and if with a good Stomach he eat them, you may give him a Quart more, and suffer him to rest till towards Noon; at what time run over those parts the Cloth covers not, with a Rubber; and having cherished him with your Hand and Voice, give him a Quart of Oats more, doing the like at one and three of the Clock in the Afternoon, or if it be in the Summer-time, you may stay till four; and when it is near Sun set, having rubbed and clad him at all points, as in the Morning, lead him forth and Air him upon hard level Ground, not too subject to Stones; Gallop him gently; Water him in due order, as has been observed, and bring him home; in like manner, when you have obliged him to stale or dung, Dress him, Cloath him, and suffer him to stand on the Bridle till such time as he has torn out of the Rack, a small bottle of Hay, at what time give him another Quart of Oats, the Manger being made clean, and leave him till about nine of the clock the same Evening; at what time coming to him again, cheer him with Hand and Voice, rouse up his Litter, and giving him another Quart of Oats, leave him to his Repose till the next Morning, and so order him every Day for the first Fortnight, daily increasing his Exercise, and keeping the Stable as dark as may be, both for his quiet and the better obliging him to feed, and by so doing you will find the advantage.

Chapter VII

Heats, what they are, and how to be Managed to the best Advantage, in Racing.

IN CONSIDERATION OF HEATS AND EXERCISE, they are somewhat different, the former being a more violent Course than the latter, and therefore twice a Week is sufficient for Heats, the days being as equally distant from each other as may be, one of them being observed to be on the Day of the Week, answering the Day whereon the Race is to be run: observing likewise not to give any Heat in rainy Weather, unless necessity compel you thereto, for it is better to vary Hours or Days than so to do; and on such emergent Occasions you must provide your Horse a Linnen-hood, made of thick Canvas, with a Bearing on the nape of the Neck, and covering the Ears, so that none but the Eyes and Nose appear: The Heats you give, in case of wholesome Weather, being to be given an Hour before it is dark, and as soon as day springs, but not in the dark, for two Causes: As, first, to prevent unwholesome Airs; and, lastly, stumbling, tripping, flipping, or falling.

The manner of Heating thus observed, the next thing to be considered is, the manner of Usage and Diet for the second Fortnight: Touching the first of these, there needs no great distinction between it, and what has been premised, only before his Bridle be put on in the Morning, you must give him about three Pints of the best Oats well sifted; after the eating of which, Dress him up and Bridle him, Cloath, Saddle, Air, Water and return him in like manner, only what Hay you give, suffer him to take it out of your Hand, and let him eat a pretty quantity if he desire it, and while he does so, let him stand upon his Bridle; as also, draw not the Bit till an Hour after; then having rubbed him all over, give him a Quart more of well-sifted Oats, and from this time forward make a Diet-bread to give him with his Oats, *&c.* according to Direction.

Chapter VIII

Bread of the first Make, and what ought to be observed in the Distribution thereof.

TAKE TO THE QUANTITY OF THREE PECKS of good Beans splent and well cleansed from the Husks, a Peck of good Wheat; mix them well, and grind them into fine Flower or Meal, and having bolted and dressed them, make them into Dough, with store of Yest and hot Water; break and tread it, that thereby it may prove the shorter, which ought to be done in a Kneading-trough; after that cover it with a warm Cloth, and set it by the Fire till it swell; then knead it again, and being well moulded, make it up into the bigness of three penny Loaves; soak them well, and when they are drawn from the Oven, turn the bottoms upwards and suffer them to cool; and at the end of three Days, parting off the

Crust, the Loaf being dry and in good order, crumb it amongst the Oats; but if it be too moist or clammy you must dry it in slices before the Fire, or suffer it to grow staler, always putting a third part of Bread, small crumbled, to the Oats: and on this quantity of Provender suffer him to rest till about eleven of the Clock, at what time renew the quantity of Bread and Oats, and leave him again till one in the Afternoon or longer, if the next Day be not his Heating-day; but if the next Day you intend to give him a Heat, then give him only a Quart of the best Oats; and when he has eaten them, put on his Bridle and tie him up, not forgetting to rub him, and do what else is convenient; at, Dressing, Airing, Watering, bringing him home and putting him into the Stable, where having let him stand a while, give him a Quart more of Oats; and then having a clean Muzzle, wash it in Beer, Ale, or White Wine, put it on to prevent his eating the Litter, or knawing the Manger or Rack-staves, if he be prone to such Vices, and let him stand till nine of the Clock at Night.

The best Muzzle for Summer is that which is made of smooth large Pack-thread or Whip cord, well knit, yet so that the Horse may freely breathe through the Lattice: and the Winter muzzle, the best is of Canvas, with a square Lattice of tape at the bottom, both of them having convenient Loops and Strings to fasten them about the nape of the Horse's Neck. The Hour of nine being come, having rubbed your Horse's Head, and other convenient parts, as have been mentioned, give him a Quart of clean sifted Oats; and when he has eaten them, put on his Muzzle, toss his Litter, and leave him to rest till the next Morning, and then coming to him, if you find him lying, give him a Quart of Oats, not disturbing him till you find him disposed to rise, which he may be obliged to by the further allurements of Provender; and these Oats you may wash in Ale or Beer, and afterward dry them by rubbing between a Cloth, and then Dress him, and putting on the Bridle and Body-cloth, lead him forth; after, by drawing up his Head to the Rack, you have obliged him to swallow a new-laid Egg, and washed his Mouth with a small quantity of Beer; and after he has dunged or staled upon the foul Litter, Rack him gently to the place you intend to Course him, suffering him, as you find him disposed, to smell at Horse-dung or Straw, if any lie in the way, which will oblige him to empty himself; and then finding him somewhat warm, take off the Cloths, and send them away, Racking him gently to the Starting-post, and beyond it, obliging him to smell to it, and then by degrees put him to a three-quarters speed, obliging him, if you find him able, to hold it throughout; that is from the Weighing-post, to the Post at the other end of the Rack: But in this case do not force him above his Strength and Wind, but bear with his yielding, which will make him take Pleasure in what he does; and so, by degrees, he will come to Perfection; observing upon what Ground he takes most delight to run, and carries his Legs best, whether moist or dry, sandy or stony, hilly or smooth, which may turn to your advantage in the Race.

Having heated your Horse, breathe him again by gently Galloping about the Field, till you find he begins to cool at what time your Servant bringing the Cloaths, put them on under some warm Hedge, to shelter him from the Wind, and ride him gently home; or you may, if he sweats much, scrape off the moisture with a broken Sword-blade, or piece of a Scithe, before you cover him, and rub him with dry Cloths, not bringing him to the Stable till you find he is thorowly dry, and moderately cool, and that he have well emptied himself, and then suffer him to stand upon his Bridle tied to the Rack; and having the following Scorwing ready prepared, give it to him in a Drenching-horn, *viz.*

Take a Pint of the best Malaga, - and pulverizing an Ounce of Per-rosin, put it therein: which being incorporated, and six Ounces of Olive-oyl, and an Ounce and a half of brown Sugar-candy beaten to a Powder, with an Ounce of Juyce of Savin or Powder of Rhubarb: heat them over a gentle Fire, and then by that means mixing them well, draw up his Head to the Rack, and oblige him to take it: by which means he will be eased of the molten grease and foulness.

Chapter IX

Other things to be taken notice of, in regard to the Running-horse, in the second Fortnight.

YOUR HORSE BEING THUS USED, great care must be taken that he catch not cold; to prevent which, you must rub him well, keep him warm clad, combing out his Mane, his Tail, and the like; keeping the Stable close, and putting large Wisps under his Cirsingles, and over them a loose Blanket or Coverlet, if the Weather be any thing cold; and let him fast two Hours or better after he has taken the Scowring, keeping him stirring or moving all the time to prevent his sleeping; which in this case he will be apt to do, and thereby prejudice himself.

Your Horse having stood the space of two Hours, &c. give him some Ears of Wheat to chew upon, and so by the extraordinary bearing and motion of his Body, the shortness of his Breath, and dulness of his Eyes, you may perceive that the Potion has met with abundance of bad Humours and gross Crudities, with which it struggles; and if so, you must forbear till the Sickness it will occasion be over; at what time take off his Bridle, turn up his Litter, and suffer him to lie down and ease himself for the space of an Hour; then rouse him, and let him tear a small Bundle of Hay out of the Rack; in the mean while dress him a Quart of Oats, and a Pint of Splent-beans; mix them together, crumbling amonst them a slice or two of the before-named Bread, and then suffer him to rest for the space of three Hours; and in the Evening, before you dress him, give him again the like quantity of Beans and Oats, clothing him up warm, but neither Saddle nor Ride him forth that Evening, nor give him any Water till the Potion has thorowly done working, only let him have another parcel of Beans and Oats washed in Beer, and so continue till Morning.

When Day appears, dress him, and let him have a Quart of Oats only; then Cloath, Saddle, and gently Ride him abroad to Air him, and let him drink, but not over-much, lest after so long a draught he swell himself: And when you return him, let him be fed with Oats, Splent-beans and crumbled Bread, in all the quantity of two Quarts, suffering him afterwards to tear a little Hay out of your Hand; And so continue him the second Fortnight, and he will increase in firmness, flesh, and strength.

Chapter X

The Running-Horse's third and fourth Fortnight; and as to Diet, Usage, and bringing the Horse to the Weighing-Post, &c.

The third and fourth Fortnight's Management being material, I shall therein come closer to Particulars: And in this case your care must be to make him a second Bread, finer than the former, but after the same manner in all respects; which being well dried and chipped, crumble it into his Oats and Beans; and that it may exceed the other in fineness, you may add as much Wheat as Beans, *viz.* Half a Bushel of each, suffering him to take his Heats on the proper Days with Ease and Pleasure, not over-straining him, thereby to render him stiff on the Racing-day.

The thing to be next observed, is, to omit the further Scowring, after you have given him other Heats; in lieu thereof, give him a Ball made after the following manner: *viz.* Take Fenugreek and Cardamum-seeds, Anniseeds, and those of Cummin, Coltsfoot and Elecampane-root, of each two Ounces: and having bruised them, and sifted off the Hulks and grosser part, add two Ounces of the Flower of Brimstone, and one of Licorish Powder: moisten them with White-wine, and add more an Ounce of the Chymical Oyl of Anniseeds, Molossus and Olive-oyl, of each half a Pint, and incorporate them with as much Flower as will make them up into Balls as big as Pullets Eggs, and keep them in a thick Glass, or well-glazed Pot for your use, they being good for a Horse on sundry other Occasions; as for Colds, Coughs, Glanders, Stoppage of the Stomach, or Shortness of Breath, in case of molten Grease or crudy Humours: And thus you may pass him over the third Fortnight with good looking to, and a due observance of Heats, Airs and Diet: From which I proceed to the fourth and last Fortnight, for so much time is sufficient to bring any Horse to Perfection.

In the fourth Fortnight, observe to let his Bread be yet finer, as allowing three Pecks of Wheat to one of Beans, reducing it to the most imaginable fineness, by bolting and other ways dressing, lighting it up with a sufficient quantity of new Ale-yest, Whites of Eggs and new Milk; working them to the best advantage, and baking them as the former, and give it him grumbled amongst the best Oats well sifted, and well rubbed between your Hands; as also Splent-beans freed from the Husks: Nor must you give him any Scowring, neither augment the Potion of his Meat. As for his Heats, the first Week you may give him two, but the second Week one is sufficient, and that some five Days before he is to run; yet, to supply the defects of the other Heat, you may give him strong Airings for the better Preservation of his Wind, and to render him the cheerfuller; as also to remove gross Fumes and Vapours, you may Morning and Evening, burn any sweet Perfume in a Chasing-dish of Wood-coals in the Stable; as *Benjamine, Storax, Frankincense,* or *Olibanum*, and often with his Oats in Ale or Beer: and the better to corroborate him, give him every other Day a new laid Egg in a Glass of Muscadel, and debar him from Hay, unless his Belly be very loose.

The last Week observe, if he be a foul Feeder, to muzzle him, unless at such times as you are with him, lest by eating his Litter, biting the Manger or Rack-staves, he injure him, or neglect his Diet; and, above all, let your Stable be so kept, that neither Pigeons, Hens, or any Fowl can come to dung in it; and the Day before he is to run, let him have his proportion of Meat in the Morning; but in the Afternoon, take away a third part, and Shooe him to advantage; as likewise dress him well, breading his Mane, Tail, and do other necessary things or ornament, but with that caution, that he may not take distaste thereat.

The Morning the Match is to be run, come to your Horse before it is well Day, and give him a Quart or three Pints of Oats, sprinkled with Muscadel or Whites of Eggs; but if he refuses to take them, then you may give him dry Oats, with a third part Wheat, well sifted and ordered, and entice him what you can to empty himself; and so putting on his Muzzle, let him remain till you have notice to bring him forth in order to run the Match.

The Warning being had, take off the Muzzle, and put on the Snaffle, it being washed in Muscadel; then rub him sleek, and cast a white Linnen cloth over him next to his Body, and over that the Horse-cloths; lay on them the Saddle well pitched with Shooe-maker's Wax, of which the Girths must likewise participate, and girt it gently, so that he may not be straitned; give him after that a mouthful or two of Oats, and pour down his Throat half a pint of Muscadel, with the Yolk of a new-laid Egg, and so draw him out of the Stable, leading him to the Course, using, by the way, your endeavor to make him empty. When he is arrived at the Weighing-post, wash his Mouth with fair Water, rub his Legs and other parts, and then uncloath him, and clap on your Saddle; then mount and wait the Signal; when starting fair, observe well your Ground, and commit the rest to the goodness of your Horse.

And thus, Reader, have I laid down the most approved Rules and Methods of this kind, which being put in practice and well observed, will doubtless turn to great Advantage, and may indifferently serve in preparing a Horse for any other Occasion or Employment.

Chapter XI

What the Rider ought to be; Backing and good Management; the Office and Duty of a Groom, &c.

As it is not every one that is fit for a Rider, relating to a Management and curious Occasions; so it ought to be considered who is, and what his chief Care must be, that he never spoil nor baulk a Horse in his first Breaking, or in Racing; for such things too frequently happen and thereby render altogether fruitless the care, pains and cost of the Breeders, and create such Vices in the Horse as will not easily be removed; and therefore your Rider must be a temperate and patient Person, not given to Fury or Anger, one of a strait upright Body, of an indifferent Weight, not too heavy nor too light, by reason the one may render young Horses, in Breaking, sway-back'd, and the other regardless of him that fits him; he must in his Function be laborious and diligent, a Lover and a Cherisher of Horses; one who rises early to Practice, and is not given to immoderate exercising: and when the Horse maketh a default, he must use his diligence to let him see it, and thereby render him tractable; for if the Person be hasty, rash and cholerick, soon provoked to impatience, he can never make a perfect Horse-man, neither can he be able to make a Horse perfect, as otherways he might. If a young Horse, of the right strain for shape, breed and colour, be well handled, he seldom fails to answer the expectation of his Master; when on the contrary, he may be spoiled, and be found good for nothing but the Plow or Cart: and this is often occasioned in a young Horse by the too much rigour of the Rider or Breaker, by which the Horse is so confounded and over-feared, that he is not capable of understanding; or else, breaking through that fear, grows restiff and sullen; when, on the contrary, sweet words and mild behavior wins so far upon a good natur'd Horse, that in a short time he will take pleasure and pride in performing the will of his Master; and, when he readily does so, or at least offers as far as his understanding will reach, he ought to be cherished and incouraged, not only with words, but with some pleasant and heartning morsel, which the Rider ought for that purpose to carry about him; at least let him spurt some Wine, or other comfortable Liquor, into his Mouth; and never correct him but when he is in a fault, always considering to strike him in a convenient place, but not about the Head, lest he dull him, or give him opportunity to be hardned in the vice of going backward; nor with your Legs and Thighs press his Rubs too hard, lest his burthen become uneasie and unpleasant; ever considering in a strait Course to keep a steady rein, not inclining to the right, nor to the left; and also an upright Body, unless in a full Course, at what time a little stooping forward, but not so far as to press your weight on the Horse's Shoulders, will not be amiss, for that will prevent gathering the Wind; the which, though it be unperceived by many is a great hindrance to the speed of the Horse: as for your Switch in such cases, it must be carried upright, or over his Head, that he may not see it to astray him; nor decline your Body from the one side to the other to oversway him, but in all thing observe a moderation and mean, being ever sure to have your ground in your Eye, and thereon manage him to the best Advantage; as also which Ground he delights to tread, which in a Race may profit much; for it has been seen that many a good Horse has been baulked by not being managed in such Ground as they naturally affect.

 As for the Office of a good Groom, it chiefly consists in keeping his Stable sweet, clean, and warm, stored with fresh Litter, by frequently renewing the old, having your Horses Cloaths, Furniture, and Materials, ever in readiness: in being expert in the convenient times and manner of Airing and Feeding; knowing what Provender, and what quantity is seasonable and sufficient; and above all to be exact in Dressing and Furnishing out for Ornament, as occasion requires, minding what Saddles and what Bridles are most fitting and convenient, and taking care to place them in due order, that they may be in a readiness when ever they are to be used: He must be a Man temperate and free from Passion, one who aims at his Master's Interest and his own Credit, such a one as has skill in Horses, and can discern the Grievances or Symptoms of Diseases, always having Instruments ready to search the Horse's Feet, or Blood him upon necessary or emergent occasions; and in case he see any imminent danger, he must not delay giving Notice thereof to his Master, or the Farrier, lest by neglect the Horse may grow past Cure; nor when a Horse is returned from a Journey heated, or the like, must not neglect him, but ever be careful to rub him well down; give him good Litter, Cloath him well, and let his Provender be proportionable and seasonable; for if a Horse be newly taken up from Grass or Soil, an excess of dry Meat will prove injurious by subjecting him to Lasks, shortness of Breath, Costiveness and Pains in the Stomach, as his suddenly being turned from hard Meat into rank Pasture will cause Laxativeness, Scowrings, Pains in the Belly, Rawness of Stomach, and offensive Rheums. And thus much may suffice for the Office of the Groom, who, as to the feeding of a Horse, may take further Directions from what has been spoken of the Dieting the Race-horse.

Chapter XII

Infallible Rules for Buying a good Horse; and how so well by sundry Marks and Tokens to distinguish Good from Bad, as well in case of Soundness as Diseases.

IN THE ELECTION OF A GOOD HORSE many things are to be observed, and especially to what end you design him when purchased; for certain it is, that no one Horse, how perfect soever, can serve to all ends and purposes. Then for the goodness of his Breed, it is very difficult to be known but by Speculation, unless you will take it upon the Credit of those that had the breeding of him, in which you may run a hazard, for few Sellers will disparage their own: However, this you may observe, that according to the Country this Distinguishment is infallible: If he be of a *Spanish* breed, then he is small and neat lim'd; if *Neapolitan*, hook-nosed; if *Dutch*, rough-legged; if *Barbary*, neat bodied and headed; if *Flemmish*, full and smooth buttocked, if *English*, well-set and strongly jointed; and so for others. Now the Colours are various, according to the constitution of the Horse's Body, and by those elected according as it likes the fancy of the Buyer, though there is difference in Goodness, to be known by the Colours, as the Horse participates more or less of the several Elementary Humours, and thus they are held:

The Coleblack, without any mixture of other coloured Heirs, is subject to Choler, and the heat thereby occasioned, inclines him to Pestilence, Feavours, Inflammation of the Liver, and other hot Diseases: To prevent which, Purges that correct the Cholerick Humours are very convenient, and Horses of this Complexion of a fierce and fiery Nature, good for War and Travel; but by reason of the Heat, which consumes the Moisture that should support them, they live not, or at least last not long.

A Horse of a bright Bay, or dark Bay, with a clear and cheerful Countenance, and a well-proportioned Mouth, is counted a lasting one, and fit for Racing, Hunting, or Travel, and is of a Sanguine Constitution; as also are the white flanked or flea bitten, white Loyard, with Hairs like Silver, or black, with a white Star, white Foot, or white-Rash; and the Diseases these are most subject to, are Consumption of the Liver, Glaunder, Leprosie, and such like foul feeding in damp weather, and the like, and will indure stronger Medicines than the former, without any danger, if they be not compounded of such Ingredients as will overheat the Blood.

A Horse that is very white, pye-bald, yellow, and the like, are Flegmatick, and fittest for Cart and Plow, or to labour in Mills, *&c.* not being capable of speed, or at least not able to hold it long; and are subject not only to lose their flesh, but to Staggers, Coughs, Catarrhs, cold Distillations, Rheumatisms, and the like Disorders, proceeding from watry humours; and therefore in case of such Disorders, hot Medicines are accounted the best and most successful.

The dark Bay, with long white hairs, russet, ash-colour, chestnut, grey or mouse dun are accounted Melanchollicks, and are subject to Inflamations of the Spleen, Dropsies, Frensies, oppression of the Heart, and pains in the Stomach; wherefore being of a dry Constitution, cold moist Medicines are requisite to be administered, and are generally heavy and lumpish; but if it happen you find a Horse with some of all of the Colours, the latter being very rare, then conclude, that the four Humours predominate by turns. And though these Observations are made for Distinction's sake, yet it is found by Experience, which is the best Master, there are good Horses, or at least tollerable good of all these kinds; therefore not any ways to discourage the Buyer, prejudice the Seller, that which follows ought duly to be considered.

If you purchase a Horse for ease and gentleness, though to Travel considerably, then the most fitting is an Ambler; then observe in the moving of his Legs, that he performs it equally, smooth, large, just and nimbly; for if he tread false, he will prove uneasie; treading short, ridds little ground; if he tread rough and uneven, he seems to roul or tumble along.

If Hunting be your aim, especially that of the Buck, or for the riding Post, or the like, wherein a large Pace is required; the running or swift Amble, as some call it, is very necessary and expedient, differing from the other Amble only in the swiftness, and is altogether easie and delightful.

As for the Trot, or as some call it, the lofty Pace, it most properly belongs to War-horses, or indeed it may indifferently serve to any other end; and those that Trot well, seldom fail to Gallop, in which case, observe that the Horse take up his Legs nimbly from the ground, yet not raising them over high, nor rowling his Body from side to side, that he beat not himself, if forcibly labored, or do it in pain, but that stretching out his fore-leg, he nimbly follow with the hinder; nor by any means cut under the Knee, which is called the Swift-cut: Further observing that he neither crosseth one Foot over or upon another, but with his far Fore-foot ever leadeth, and not with the other. If these Qualities are found, then the Horse is good for speed; but if he gallop roundly, raising strongly his Fore-feet, he is fittest for a Charging-horse, or one to be trusted in carrying his Master through a dangerous Attack. If his Gallop be slow, so it prove sure, he may make a Traveller; but if he handle his Legs confusedly or shuffingly, then be aware of some defect, especially if he Gallop painfully.

As for the Stature, that I leave to every Man's discretion, as his occasion sutes or requires; the largest being for strong solid labour; the middle size for swiftness and long journeys, if not over-burthened, and the lesser size for ease and moderate recreation: And what is more remains materially to be known of this kind, is the Deformities and Imperfections, that is, how to discover them; and these are contained in natural and accidental Deformities, inward and outward; many of them hidden, and so secretly couched, insomuch as they will deceive the Skilful: And of these, especially of such as are the most dangerous, and materially to be known, I shall acquaint you:

First, observe and inform your self what you can, as to the Breed and Paces of the Horse; when, having taken off the Saddle and Bridle, leaving a Halter only upon his Head, cause him to be rubbed down well, and stand just before him, earnestly beholding his Countenance, whether it be cheerful and sprightly, free from cloudiness or scowling; that his Ears be small, sharp, thin and pliable; and tho' they be inclining to longness, yet if they be well proportioned and well set on, it is not much amiss, for his freeness, beauty, and goodness of Mettle are thereby declared; when, on the contrary, if they be loleing, thick, heavy, dull and clouterly, weakly set on, and seldom moving, dullness and a heavy Disposition are denoted. If the Forehead of the Horse swell, and the feather or mark stand high, above his Eyes, or on top of his Eyes; if it happen he have a white Star, or white Blaze, evenly placed, of an indifferent size; or if it be a Snip on the Nose, it is a mark of Beauty, but if his Face be flat and cloudy, they are Tokens of dullness and defect; if the Star or Ranch be awry, or low, it is unseemly; or the Nose instead of a Snip, be Bald, it signified badness, and a Horse not fit for any considerable Service; but if the Eyes be brown, black, shining, staring or standing out, and move with an equal Motion, the black of the Eye filling the fleshy Orb, so that little or nothing of the White appear, then it is a sign of goodness, and a well-mettled Horse; but on the contrary, if the Eyes move slow and uneven, and the White greatly appear; if they seem cloudy or dull, then it is a sign of badness, and the like; if they seem wrinkled, or are very small; if they be of a fiery red, they incline to Moon-blind or Moon-eyes, which are but a remove from Blindness; if the Eyes be white and walled, it denotes a weak sight; if very bloody, inclining to blackness, bruises are thereby signified; red, dull Eyes, and a hollowness, fore-run Blindness; if the Eyes matter, and much rheum flow from thence, then the Horse has been extremely abused by over-riding, or else is very old; and so of the rest.

When you come to handle the Horse's Cheek or Chaps, if there you find the Bone thin and lean, the Wind pipes or Throple full and large, even, and free from knobs and kernels, and the Jaws standing even, so that the Teeth shut like a Box, and above so placed, that the Neck may seem to shut or sink into them, it is a sign of a free Horse, and one of a good wind and courage; but, on the contrary, if his Chaps and Cheeks be thick, fat and suddenly inclined, the throple small and muffled up with a thick gross substance of fat, (then especially if it appear much kernally) it not only signifies short Wind, but threatens the Glaunders, Stargles, and other Diseases incident to Straitness and Pursiveness; and most particularly dangerous Colds of many Natures.

If it happen the Jaws to be so strait, that the Neck swell over them as it were, yet if it be occasioned by no Distemper, it matters not much as to the service of the Horse, though it is very unseemly; but if it swell out long, and end taper, then the Horse may be subject to the Vives, Tumours, Imposthumes, or the like: Then observe that his Nostrils be wide, large, open and dry, and that the redness naturally appears without any forcing or straining: If the Lips are even, the Mouth deep, and the muzzle small, they signifie good wind, speed and courage; but if on the contrary, dullness and infirmity; as also by the shortness of the upper Lip, and wrinkles on the side of the mouth old age; and above all, observe his Teeth be clean, even, firm and well proportioned. But of them more particularly hereafter, when I come to speak of the Horse's Age, though, in general, unseemly and even Teeth are a sign of badness.

Observe the Breast of your Horse; and if it appear broad, bowing outward and well feather'd, it denotes health and strength; when on the contrary, a small Breast, ridged or flat; as also the Shoulders standing in, denotes stoppage, weakness in travel, pains of the Heart and Liver, and the like, with stumbling and interfering; as also does a Breast that is narrow and imbowed inward; those being generally the marks of a Horse weak, and unfit for labour.

Next cast your Eyes from the Elbow to the Knees, and see his fore Thighs be well proportioned, strait, and well clad with Flesh and Sinews; by which strength is signified, and by the contrary, weakness.

Observe the Knees, that they are carried even, that they be well jointed, close knit and fixed, with Sinews and Ligaments free from Scabs and Scars, not over round, but well jointed, and boned; for if they appear contrary to these Marks, as round, and swelling of a more than ordinary bigness; if they be scabbed, broken, or the Hair off, then it betokens an uneven treader, and one that is apt to stumble, or subject to the Swift-cut.

Having thus far taken notice, the next thing to be observed materially, is from the Knee to his Pasterns; and there consider, if the Legs be clean, well fleshed and well sinewed, bowing somewhat inward, which shews strength and a firm treader; but if on the inside you find any fleshy Excressences, or Scabs, a little beneath the Knee on the inside, likewise then is the Horse subject to interfering; but if a general Scab, &c. Then the cause is foul keeping; if on the inner bowing of the Knee you perceive Seams, or the Hair broke disorderly, a cankerous Mallender is denoted, or Ulcer.

Observe that the Pasterns be clean and well knit together, especially the first, and the other short and strong, standing upright; if the first be large or swelled, then if the Horse frequently subject to the Sinew-strain, and twitching girds; if the other be weak or bending, it will hardly support the Body in any long Journey, or uneasie way.

As for a good Hoof, it ought to be black, rough, and much inclining to roundness; for a long, white, rough, or brittle Hoof, denotes an uneven tread, a tripping or losing the shooe upon every occasion, and is altogether unseemly, and fore-shews the Horse subject to foundring. As for the Crownet of the Hoof, observe if the Flesh swell a little, and that the Hair lie decently in good order, that no Scabs nor boney Excressences happen there; for if those appear, ten to one but the Ringbone will afflict it.

Thus having observed your Horse in the front, place your self on the right side of him, and observe that his Head be neither too high nor too low; that his Neck at the setting on be small, and indifferently long, growing deeper by degrees, till it arrive at the Shoulder; then for the Crest let it be high, strong, and somewhat thin; his Mane somewhat curling, thin, long and soft. These are not only Signs of a good Horse, but a Beauty also; and consequently the contrary of Deformity.

In the next place, have a regard to the Chine of his Back, that he have a due propoertion of even, broad, and strait; his Ribs somewhat large, and bending outwards like a Bow; his Fillets short, strong and upright, about four Fingers distance from his last Rib and his

Huckle-bone; his Body being well let down, yet hidden as it were within his Ribs; his Stones round and even hanging: and these are Marks of Beauty and Perfection. Whereas on the contrary, a narrow Chine will subject the Saddle to wound the Horse, and bending or saddle-back'd denotes weakness; to have the Ribs over-fat, makes the Horse breathe painfully, when hard labored. If his Fillets hang long and thin, they denote him weak, and not capable of performing a Journey in a hilly Country; and if his Stones hang down, as in long Bags or Purses, uneven and unseemly, it denotes a defect in Nature, occasioned either by Sickness, or a feeble Constitution.

As for his Buttocks, note that they be round, full, and plump, and that the Tail be well set on; the best observance being the evenness of the Buttocks with the Body; and although they are something long, that they spread well behind, and are not inclining to make a ridge of the Crupper-bone, nor stand long and narrow; for in such caves neither a Pillion nor Pad-saddle can sit easie nor convenient.

The hinder Thighs commonly called the Gascoins, are likewise to be regarded; and the observation is, that they be well let down in an evenness to the middle joynt, thick, brawny, full, and swelling, by which strength and goodness are to be observed; whereas a lean lank Thigh not well grisled, but very slender, denotes weakness: Then observe the middle joynt behind, and see that it be somewhat lean, well knit and ligamented, and Sinews well veined, and moderately bending, which shews Perfection in that Part, but if it have any Chops or Sores in the inward bending, then it threatens a Sellender; and if it have a general Swelling, then some blow or bruise has happened; again if the Swelling be particular in the Pot or of hollow Part, or on the inside, and there it happen the Vein be full and proud, and the Swelling moreover short, then is it the Blood-spavin; if hard the Bone-spavin; but if it happen behind, on the Knuckle, then is it a Kerb or Kurb.

After this look on the hinder Legs, observing whether they are clean, well fleshed, and supported with Sinews: if so, it is a good sign; but, if they be fat, they will not endure Labour; and if they swell, it is a sign the Grease is molten in the body of the Horse, and that he is foul; if scabbed above the Pasterns, then beware the Scratches; if under the Pasterns, Chaps appear, then it is the Pains: All which are infectious and dangerous to the Horse, and render him far from being good.

As for the Tail, it seldom fits amiss upon a good pair of Buttocks; and on the contrary, never well upon ill ones; but the best setting on is broad, high and flat, a little couched inward.

Chapter XIII

Infallible Rules to know the Age of a Horse by his Teeth, in all their particular and general Marks; as also by the Tail.

The full number of a Horse's Teeth are computed Forty, and seldom, when he has all that he will have, prove less: And thus they are distinguished, *viz.* On either side, above and beneath, six great wang Teeth, which compute at twenty four, generally termed Grinders; then six above, and the like number below, in the forepart of his Mouth, which, with the other, make Thirty six, these latter being properly called the Geatherers: and, moreover four Tushes, two above, and the like number below, generally called Bite-teeth: And from these the Considerations are as follow:

In the first Year only the Grinders and Geatherers, commonly called Colts-teeth, only appear small, white and smooth.

The second Year two of these Teeth are changed in the middle of the Geatherers, *viz.* one above, and another below and as for those that come up in the room of the shead ones, they are generally darker or browner, though they may be distinguished other ways by the largeness.

The third Year, next to those before-mentioned, are changed, leaving the rest to be changed the successive Years, so that after the fifth Year no Colts-teeth, and then likewise his Tushes are compleat; and those that succeed the last Foal's Teeth will be hollow, having in the midst of them small black Specks, of which many take singular notice, shaping their Judgment thereby; though if they take not heed, they may be imposed upon by artificial as well as natural Specks or Spots, though indeed by these natural Spots a Horse's Age may be known till eight Years are past, and no longer.

The sixth Year being come, his new Tushes arise, at the bottom of which young Flesh will appear, the Tushes being small, white, short and sharp, and so continue to the seventh Year; in which Year they are of full growth, and the Mark very plain to be seen and observed.

The eighth Year all the Teeth will appear plain and smooth, but the Specks, generally called the Marks, will decrease by degrees, till they are seen no more, ending as it were with or before this Year, nor will the Tushes continue longer white, but incline to a yellowish. The ninth Year being come, and the Mark consequently no more to be seen; consider the Fore-teeth will be broader and longer than usual, their colour inclining to a yellowish paleness, and the Tushes to bluntness.

The tenth Year's Age may be known by the filling up of the holes within side of the Tushes, which were before like Cells or hollow Pits; as also his Temples will be crooked or distorted.

The Horse being arrived at the eleventh Year of his Age, the most material thing to be known by the Teeth is their distortion, yellow colour and unseemliness; as, also their more than ordinary length. Now as for the twelfth, and generally held to be the last Year, a Horse's Age can be known by the Teeth; your observation must be taken from their largeness, foulness, and hanging over each other; and though some have held, that by the close wearing of his Tushes to the chips, his age may be further known; yet let me advise the Buyer, that he may not be deceived therein, that being no certain Rule or Mark, especially to rely on. And thus much as to what relates to the knowing a Horse's Age by the Teeth; from whence I proceed to his Mouth in general; and that which I infer from thence is this:

If upon having regard to a Horse you perceive his Teeth and Lips hang over, uneven or unpleasant, unless it proceed from some Distemper in the Head, or Disorder in the Body, then is that Horse old.

Observe you find not too deep a burning for the *Lampas*, that so the Provender may be apt to stick therein to his hindrance in feeding; and that the spaces between the Barrs are not too deep and hollow of themselves, for that denotes Age, but that they be pleasantly fixed, smooth and soft.

As the Eyes are, if sprightly and well fixed, a great ornament to the Horse, so if other ways, they render him uncomely and deformed; wherefore the defects ought to be considered, *viz.* If upon view you perceive the Eyes to be sunk, dim, hollow-pitted, the Temples crooked or wrinkled, then old Age is signified: But if the contrary, *viz.* If the Pits be full, the Eyes lively, bright and sparkling, the Temples strait and smooth, then is the Horse young.

If you mistrust a Horse's Age, and cannot well find it according to the former Directions, then apply your feeling to the Tail, and if there you find two knobbs near the Setting-on, each about the bigness of a Musket-bullet, then you may conclude your Horse to be very young; but if the Joynt thereabout be plain and smooth, no such sign or token appearing, then conclude him to have passed the tenth Year; and so by those knobs, [proportional to the bigness, consider his Years.]

Other Considerations there are to be had from the Hair and Skin, as, if you perceive any Hairs that are grisly about the Eye-brows, especially if the Horse be of a dark of contrary colour; or if such be found underneath, the main Age thereby is signified: The like if a white Horse, or one very light coloured, happen to have black or red Mannals. As to what is to be observed from the Skin, let this be a Rule: if the Horse be old, the Skin you raise between your Thumb and Finger will stand in a pucker, and not suddenly return to its Place; but if young, it will fall flat almost as soon as let go. Nor is the Hoof to be altogether neglected in this case; for if you cast your Eyes upon a Horse whose Hoofs are crinkled, seamed, rough or unseemly; or if they be brittle, and subject to break or crumble, then is the Horse old or infirm. And thus much may suffice as to Age, unless somewhat that will occur by the way in handling other Matters of the like nature.

Chapter XIV

How exactly to know the state or Condition of a Horse's Body relating to Fatness or Leanness, Health or Sickness.

IF YOU FIND YOUR HORSE IN A DROOPING CONDITION, if he refuse his Meat, and be heavy or dull eyed, seem uneasie, standing, or restless, lying down, then conclude some inward grief or defect afflicts him, not to be so easily discovered as some have conjectured: Wherefore I shall first point my Directions at the State of a Horse's Body in general, and from thence to particulars: beginning, the better to give the Reader an Insight into the rest, with the strong Estate, or healthful Constitution of a Horse, &c.

As for some Horses, though they are in Health and good Temper, yet by reason of their roundness and well compacted Body, the Buyer may have a lean one put upon him instead of a fat one, unless he be skilful in handling and trying, thereby to discover his want of Flesh, or feeble Constitution; and so a Horse that is long and full made, may seem lean to the Eye, at a distance, when really he is in good case; or a fat Horse may likewise deceive the expectation of the Buyer, by having contracted inward fatness and foulness, which will require time and labour to avoid and bring away; when on the contrary, some Horses will be fat without, and in good case, yet be clean within, and be free from gross Humours; and others again, though outwardly appearing lean, will be inwardly gross and foul: and these proceed from the Order and Constitution of the Horse &c. And this is to be known upon search by demonstration, sundry ways, as to what is outwardly to be taken notice of.

First feel the Horse's Body in all the advantageous places, but, especially upon the Ribs; and in so doing, if you perceive a general Softness of the Flesh, or looseness, so that your Fingers easily sink into it, then conclude a foulness within; or if it be hard and firm upon all but the hindmost Ribs, then suspect Molten-grease within, which must be voided by Purging or Scowring; though at the same time the Horse appear poor and lean, or else he cannot well be reduced to a good state of Body; and the better to make the Potions operate, is to prepare him for them by Exercise, that being moved and stirred, it may the easier be brought away,

If so it happen, as has in some manner been said, that the Horse appear to fat and thick, that the space between his Chaps is as it were closed up, or the Jaws upon handling prove fleshy and full, then much foulness in Head and Body is signified; but if upon handling he happen to be thin and clean, though some hard knobs be contracted, yet that may rather proceed from a Cold taken, than from Foulness, &c. But to come to Particulars:

If the Horse's Dung appear of a pale, whitish, yellow Colour, not too hard, yet so that in falling it only flat a little, but does not break in sunder or crumble, and be not of a rank noisome scent, then is the

Body of the Horse free from inward foulness: but on the contrary, if it be of a black or muddy Colour, hard, like Pellets, yet hot and greasie, it betokens foul feeding, and a neglect in the Keeper of not giving him a due proportion of Corn, but rather feeding him with Hay in over-abundance, or suffering him to feed foul by eating his Litter, denoting he has likewise Molten-grease within him that cannot come away without some Artificial Helps. If the Dung be of a reddish colour, and hard, it signifies the Horse has been over-heated and over-strained, and therefore ought to have a Glister of cooling Simples to open and cool his Body, thereby preventing Sickness, or at least Costiveness.

If the Urine be of an Ambor colour, somewhat inclining to thickness, or a strong, yet not offensive scent, then it denotes a healthful Constitution; but if it be ruddy or high-coloured, hot and scalding, so that the Horse makes Water with Pain, or does it by degrees, then has he taken hurt by being over-heated, or in being ridden too soon upon being taken up from Grass: if so it happen that the Urine be of a high colour, and yet very clear, like Beer, well settled, then it is occasioned by Heat or Inflammation, caused by a Surfeit.

If a white film or scum happen to be upon the Urine after settlement, it denotes the Horse weakened by too much Venery; and in a Mare by too often Breeding, too furious a Lust, and consequently not very fit for Service, by reason of the weakness of his Back. Green Urine denotes a Consumption of the Body; and that with bloody streaks or specks, not much unlike to a mattery Substance, denotes an Ulcer, or the like defect in the Kidneys: and, further, a lead-coloured, gross and cloudy Urine, denotes a dangerous Sickness, if not Death itself. And thus much as to what is to be known in relation to Health and Sickness by these Excrements.

And now let us consider a little, if a Horse be out of order, how those that attend him ought to carry themselves to him, in relation to his Diet:

Let him observe in this case the Disposition of the Horse, and not violently thrust or intrude the Provender upon him, but consider to give him what he best likes, and that at leisure and by degrees, winning him with mildness and sweet behavior; and be wary, above all things, not to give him any offence, increasing the quantity as you perceive him desirous to take it, giving him white Water to drink, and ply him as you see occasion, with warm Mashes, warming his Water with a little Bran dusted over it, if the Weather be cold; and be sure to suffer him to be in a warm Stable, with clean Litter and good Dressing.

If your Horse be any ways subject to Lameness, bewere in your Coursings or Heats you avoid craggy and stony Ground, or forcing him too much against a Hill.

Chapter XV

Observations to be taken as to the State of a Horse's Body from the Privy-parts, Limbs, Feeding and other matters, &c.

The Privy-parts of a Horse, as to Sickness and Health, are much to be observed; and this especial notice to be taken, *viz.* That if he be a Stone-horse, his Cods hang even, and his Stones well trussed up, firm and round, which shews him in good Heart, and fit for Business; when, on the contrary, if they hang long, or bagging down, a weak Constitution is thereby implied: Nor is it less to be observed by the Yard of a Gelding, or shape of a Mare. But, for Brevity's sake, waving further particulars, I shall proceed to take Observations from the Limbs, and what, in case of defect, ought to be done on that occasion:

In case you intend to Run or Travel your Horse, if you find his Limbs stiff, or not so pliable as you would have 'em, or as ordinarily as they happen to be; then to render them supple and pliant, Melt Hog's Grease and Oil of Chamomile, and dipping a Woolen-cloth hot into the Liquids, dipping a Woolen-cloth hot into the Liquids, anoint or bathe therewith his Legs or Thighs from above the Knee, and from the Cambrels downwards; or, for want of these, Neat's Foot Oil; after which chafe it in with your Hands; and in often so doing in all places where you see occasion, you will wonderfully restore his Limbs, rendering him supple and pliable.

There are many things to be observed from the Sweating of a Horse, especially from that which proceeds without external force, occasioned through hard Labour, or the like; for then is the Horse faint, foul fed, or wanteth Exercise to evaporate the abundance of watery Humours which he has contracted: and especially this is to be noted, when his Sweat is white and frothy, like a Lather of Soap; but in case of Heats, and other hard Labour, to create Sweat; if it appear black and pearly, like clear Water, then is the Horse in a good plight or habit of Body, especially if he be lively and brisk, not in the least afflicted with any shaking or trembling; for that infers the Sweat forced out by some internal disorder, rather than naturally evaporated.

Chapter XVI

Of the Elementary Parts of a Horse's Body, and of the Agreement of the Humours therewith. A Discourse of Corruption and Generation, as to Goodness or Badness, Health or Sickness, &c.

As for the Body of a good Horse, it differs not in the Elementary Compostition from that of Man; for of all the Elements it consists, *viz.* Fire, Air, Earth, and Water, which indeed are the Generatives and Consumers of all mortal Things and Beings; and although these Elements are different in themselves, yet are frequently agreeable in Composition; for the Fire, though hot and dry in it self, yet compounded with the other Elements is a mortal Frame, or bodily Composition, diffuses a gentle Heat to nourish and support Life. Though the Air is hot and moist, but more participating of the latter, yet does it temperate the Heat of the former, and gives a kind of respiration and gentle breathing to refresh and exhilarate each Part and Practice. The Earth, though it is in it self gross and ponderous, yet in this Composition is it rarified, and by its substantial matter composes the Frame wherein the rest of the Elements cohabit; and being cold and dry, serves to temperate the hot and moist. The Water, though cold and moist in it self, serves in this case to moisten and render pliable the Sublunar Works of Nature, and enables them to subsist: And all these four Mothers of Creation participate more immediately of the four Humours, of which the Bodies of Animal and Rational Creatures are composed, as Choler, Blood, Melancholy, and Phlegm; the first of Fire, the second of Air, the third of Earth, and the fourth of Water: and as these more or less operate, so according to their qualities, is the Body moved and exposed to heat, cold, moisture, or driness, and consequently to the effects they produce; for these have their principal Dominion in the Seats of Life, and from thence extend their power and force to every part and member. As this Phlegm possesses the Brain, Choler and Blood, the Liver and Heart, and Melancholy the Spleen, which is the Receptacle and Conveyance of the Excrements of the Liver, all of them having distinctly and conjunctly their proper Office assigned: as thus, the Blood is the principal Nourisher of the natural Frame, Phlegm or Moisture renders the Members flexible and compliant in motion and use. Choler causeth Digestion by its operative Heat; and Melancholy disposes to an Appetite, and attracteth the grosser parts of nutrimental Elements, and occasioneth the disburthening of Nature.

It is generally agreed upon by the Learned, That every Organical Body is supported by four principal instrumental Members; and these are held to be the *Brain, Heart, Liver,* and *Genitals*, these performing their particular Offices and Functions: For as the Sinews are supported by the Brain, the seat of Animal Spirits; the Arteries from the Heart, or seat of Vital Spirits; the Veins, which are natural Parts, from the Liver; so the Seminal Vessels are supported by the Genital Parts of instruments of Generation; and these conjoynly operating, are the Elementary Substitutes, as participating of their Nature, and consequently the Materials of Generation.

Having briefly discoursed of the Elementary Parts of the Bodies of creatures, I now proceed from the four Humours, Elements, and Instrumental Members, to the Natural Faculties; which in this case are the next things to be considered, which are, Retaining, Concocting, and Expelling; and though all the Humours are Instrumental in promoting these, yet principally Nature serves herself but of only one to work upon, and that is a Wheyish kind of Blood, generated in the Liver, or attracted thereby from the purest part of such Nutriment as has been concocted in the Stomach, and from thence conveying it self to the Liver; and through the great Vein, conveying Nutriment into all its branches, and consequently into all the parts of the Body, by dispersing the rarified Blood into every part, which, by the help of the other Humours, supports the frame of Nature. As for the watry part of Nourishment, and that of the grosser substance; the one is carried into the Bladder, and the other passing into the Bowels, are in the end cast out of the Body to make room for more of the like nature.

But, moreover, there are two Veins that conduct part of the purest and rarified Matter into the Seminal Vessels, where, by the operation and contraction of the Generative Parts, it is refined by a gentle heat to a more spirituous quality, and so in the end become seed, which, according to the goodness or badness mixed in the Womb, proves effectual or ineffectual: for Note, That if the thin and subtil Blood be wanting to support the Seed, and enliven it with heat, it frequently fails in performing its Office, and the Horse becomes fridged and uncapable of performance, and is often subject to moist Diseases; as the Glaunders often proceeding from the Liver and Lungs when they are infected with Putrefaction, occasioned by moist Rheums, and other indigested Moistures descending on them, nor Inflammations occasioned by Lumps, Knots or Kernels under the Chaul, proceeding from cold or immoderate Labour, and many times the Morning of the Chine by a Horse's immoderate Leaping a Mare, or over straining himself in the Action: And this may likewise happen by eating too much raw Meat, or drinking upon a sudden Heat, and many other ways; which, when I come to treat of Diseases, I shall largely Discourse on. But first I shall proceed to let you know how Horses ought to be used, to prevent their contracting Diseases as much as in the superficial Schemes of Art the greatest Artists of this kind have allowed; and the means are cheafly five, as Cleansing, Blood-letting, Purging, Sweating, and Vomiting.

The first of these may be said to be two-fold, as outward and inward; the first being by cleansing his external Parts by Care and good Dressing, which ought the more diligently to be observed upon his being taken up from Grass; and the time limited for that, by the Curious, is *Bartholomew* tide, for the heart of the Grass beginneth to decline. And this may be done in the manner which I have often recited in what relates to Dressing; nor is it at all inconsistent with Reason, that the cleansing the Skin from dust and dirt, and loosening it in every part by gentle rubbing, should enliven the Horse, and render him more sprightly; so that Nature having her free course and progress without restraint, may operate in dispersing kind refreshment to every part, and keep those gross Humours from settling, that are frequently the original of Diseases and Grievances; and so if it happen that your Horse be miery and foul, then you may use Fulling-earth, Soap, and other scowring Materials, especially with warm Water; but then having an especial Care he not catch cold thereupon, keeping him warm, and well drying him till he be thoroughly cold; and in so doing you will prevent those moist Diseases wherewith he is always afflicted; and the better to effect this, cut away all the superfluous Hairs that any way annoy the Body, or other Parts, *&c.* And so being shooed and neatly dressed, I leave him to the second Part, which consists in the Internal cleansing and purging Directions; for which, in consideration of more immediate or particular Directions, take in the following Chapter.

Chapter XVII

How a Horse ought to be used in general and particular, as to his Physick, Diet, and Looking to.

As for the inward Purging, the measures ought to be taken from the temperature of the Horse's Body, and more especially in case of his retirement at such a time, that no hard Labour or immoderate Exercise has been imposed on him: and in this case if you find him Costive, or that in case of Evacuation, Nature help not, as in usual cases, let some one with a small Arm penetrate his Fundament, and draw thence the Dung that obstructs, or at least clogs the fluent or natural passage, that so both the natural and artificial Motions and Causes proceeding from the Drugs, or composition of Purgation, may operate and perfect the intended design, and may be best administered Clyster-wife; but then, consider the constitution of your Horse; if he be fat, and somewhat inclining to foulness, it must be a strong Potion that will effectually operate; but not so if the Horse be weak and melancholy. But, waving 'em here, I shall speak more particularly of them in due pace; as also of Drinks which ought to be made, and seasonably given for the preventing Sickness, and preservation of Health; and if you find the Blood coagulated, which may be perceived by trembling of the Veins, and the working of it therein, then it is requisite to let him Blood, and give him a gentle Vomit to carry the foulness from off the Stomach, that may be the occasion of bad Digestion, and consequently the naughtiness of the Blood; Dieting him with Marshes and fine Provender, from which may spring such wholesome Nutriment, as may create a thin and airy

Blood; rubbing and often anointing his Body with Hog's Lard, or Ointment of Mashmallows. And not to be satisfied what things mostly contribute to Health, take the following Opinions of the Learned, *viz.* A good natural constitution, good digestion, good nourishment, moderation in feeding and diet, moderation in labour and sleeping, and moderation in leaping of Mares. Again, wholesome airs, not laboured too soon over Grass, to be kept from raw Meats, not to drink nor eat being hot, ever observing to walk him at the end of any Journy, and not to Physick him unless you find occasion. And these Observances being the occasion of long Life, I shall endeavor, for the better satisfaction of the Reader, more fully to demonstrate them.

As for Nature, good Digestion, and sound Nutriment, they ought to be constant, and indeed they are well proportioned, when neither the Moisture with its Humidity is not so predominant as to quench and over-power the Heat, nor though of necessity the latter must be of force above the former, or else Digestion cannot be perfected as it ought; and in that case seasonable Nourishment must consequently fail.

Moderation in Eating is another main cause of long Life, as immoderate Eating is of a short one: For as excess in Eating, though the Provender be never so good and wholesome, obstructs good Digestion, and contracts crudities with a bad habit of Stomach; so on the contrary, spare Diet weakens and decays Nature, and infeebles the natural powers and faculties of Life, giving the Heat by that means a power absolutely to subdue and conquer the Radical Moisture, and gives Diseases an opportunity to break in upon the infeebled Body, which prove many times too hard for the Farrier; for in all Creatures observe, that the weaker Nature is, the stronger is the Distemper.

Another cause of Health and long Life proceeds from moderate and kind Labour, for by indifferent motion digestion is much forwarded, and the Humours dispersed, being by that means prevented from settling more than is requisite in any one place; besides, it is the cause that Excrements are sooner voided, which by lying down in the Body might occasion Sickness. And further observe, that he be not laboured upon a full Stomach, so as by over straining digestion may be hindred, which should turn to Nutriment, and lay a foundation for Diseases by indigested Crudities, being too suddenly drawn into the Veins, and by that means dispersed into all parts of the Body.

Another cause there is of the like nature that depends upon the moderateness of sleeping and waking, for too much waking is an enemy to Health, by spending the vital Spirits that should support and maintain Life, and a decaying that Moisture that should refresh the several parts of the Body, causing thereby leanness and barenness, a dulling of the Brain, and a defect in the Lungs and Liver, whose Offices it weakens by decaying the vigor in the performance; and contrary to this, excessive sleeping dozes the Brain, hinders Digestion, and obstructs Nature in the performance of her Offices, contracting noxious Vapours and a Foulness of the Stomach.

Another thing to be considered is, that your Horse be not admitted to spend himself too much upon Mares, because such immoderate Exercise weakens the Brain, Back and Eyes, wastes the Vital Spirits; when as gross Air or evil Scents not only make the Horse loath his Provender, but corrupt the Blood, and subject the whole Body to Diseases.

Travelling after Grass too soon, without purging and cleansing the Horse's Body, causes the bad Humours to incorporate, or by spreading themselves to afflict each part with Pains and Disorders, reducing the Horse to a dullness of Temper and Disposition, and so raw Meats engender raw phlegmatick Humours, afflicting the Stomach and Brain, occasioning the Glaunders, Coughs, Catarrhs, Stavers, Yellows, Anticors and Morfoundring, not only disable the Horse, but if neglected, become incurable but by Death.

Another cause there is, and that not to be lightly regarded, which is, not to suffer your Horse to eat or drink when hot, and to stand thereon, for by so doing the Blood will corrupt and putrifie, occasioning Surfeits, Feavours, Obstructions, and many the like Maladies and Grievances, frequently occasioning Death: But as soon as you bring him home in that condition, put him into a warm Stable without washing; rub, or cause him to be well rubbed down; cloath him well, and let him have a sufficient quantity of warm Litter to stand on; and if he be subject to eat it, put on his Muzzle and so let him stand an Hour or more till his Grease be sufficiently cooled, and you'll find him in a fit condition to give him his Provender: And, lastly, that he may be well breathed, and sound winded, thereby being enabled to hold out as occasion shall require, you may at seasonable times mix with his Provender the Powder of these following Seeds and Drugs, *&c. viz.* Cummin and Anniseeds, Powder of Licoras and Elecampane-roots, Barberries, Brimstone-flower, and the Roots of white Lillies, Hysop, Horehound, Savin, Colts-foot, and the Seeds of Marshmallows, Rue, and *Polipodium* of the Oak, and many of the like quality that will occasion good Wind, and prevent Infections, purifie the Blood, and help the Liver in the performance of its Office; that is, to rarifie Nutriment by a good Digestion; but you must not mix over much at a time, especially when you first begin it, lest he take disgust at the scent or taste, and so reject his Provender; but increase the quantity by degrees, not giving him any above twice a Week, and then let it be Morning and Evening.

Chapter XVIII

An exact Description of the Veins of a Horse, how situate in the Body; as also of Blood-letting; and how, and upon what account of Sickness or other defect, they are to be opened, for the prevention of Death or Danger.

From the Liver, note, there ariseth one large Vein, which, like a Conduit, supplies the rest of the Branches, which are many in number, and spread themselves throughout the Body like little Rivolets or Streams: And of these, two material ones are found in the Palate of the Mouth, above the first and third Barrs, which ought to be opened by a discreet Farrier, when the Horse is afflicted with any Malignant Pain, or Disease in the Head or Stomach. Two more there are that are descending, which, from the lower part of the Eyes, descend to the Nostrils, and are best opened when the Eyes are afflicted with any Distemper or Grievance. Two others there are above his Eyes, which are called Temple-veins, because they run cross the Temples; and these are generally opened for cold Diseases in the Head. Two great ones there are likewise that run along the Wind-pipe, by the sides of it from the uttermost Joynt of his Chaps, to the Breast, commonly called Neck-veins; and these are opened for sundry Diseases, being the most useful Veins that are opened. There are two other Veins that arise from between the Fore legs, and are called Breast-veins, because they end on top of the Breast; and these are opened in case of Surfeits, Feavors, or Heart's Sickness. Two others there are which ascend from the Fore-legs, but rise not so high as those before mentioned; and these rest upon the foremost Bough of the Fore-leg, and are generally called Plat-veins; and are opened in case of Splents, Spavins, Mallenders, or Sallenders, &c. Then are there four Veins which run along the Fetlocks of the Horse, known by the name of the Shackle-veins; and these, although they are small, are yet important ones, and by bleeding cure the stiffness of the Joynts, and prevent Foundering. Then are there four more about the Corronets in his hoofs, called Corronet-veins, and are opened for the Ring-bone and Surbating. In the Hoof there are four more, which circle his Toes, called Toe-veins; and are opened for Fretting and Foundering. Two great Veins there are, that, descending from his Stones, pass along the inside of his Thighs, to his Cambrils; and these are called Kidney-veins, and are opened with success for Diseases in the Reins and Kidneys. Two others there are, which, descending from above the hinder Cambrils, pass along on the inside of the Hinder-legs down to the Fetlocks, being called the Spavin-veins, and are seldom opened, unless in case of the Blood-spavin. Two Veins are likewise found in his Flanks, from whence they take their denomination, and are opened successfully for any Pain or Grief in the Fillets or Reins. Two Veins he has in his Haunches, called by the name of Haunch-veins, and are opened in case of falling away, or consumption of the Flesh, Hideboundness, or the like. Two there are that run along the side of the Belly, from the Elbow, to the Flank, and are called Spur-veins; and in case of Fretting, Belly-foundring, Spur-gall, Festring, or Swelling of the Belly or Flank, they are opened with success. In his Tail is one single Vein, called the Tail-vein, which is opened in case of sheading the Hair: So that in all there are found, as a compleat number of principal Veins, or as from which Blood is taken on sundry occasions, thirty-seven; in knowing which, any Man may understand how to let his Horse blood, in case of the before mentioned Distempers, at such times as a Farrier is not at hand. Many other small Veins there are, but so inconsiderable to our purpose, that I shall not undertake them in this place, but proceed to speak somewhat of the Sinews, which is another Material part of this Noble Creature, worthy to be inspected.

Chapter XIX

A Discourse of the Sinews and their Situation, with their Use and Office; and what in that kind is to be observed as to the State of a Horse's Body, &c. with a Description of the Bones, &c.

As for the Fountain or Source from whence the Sinews spring, and by which they are nourished, it is the Brain, for there passes through the hollow of the Neck one large Tendon or Sinew, which running along the Back-bone, continues its course even to the end of the Tail: And from this proceeds two small Branches, which passing through the Scull, fall down along the Horse's Cheeks to the points of the Nostrils. Two other Branches there are which pass through holes in the nether Chap, knitting it with the other, and so pass along by the great Teeth, meeting just beneath the nether Lip. Then are there twenty-eight small Strings, that running through the seven Bones that compose the Neck, knit them together; and, passing from thence, use their office in knitting; likewise the Chine even to the Strunt. Then are there two great ones that stretch themselves over the Spade-bones, and then dividing into divers Branches, spread in a descending manner into the Legs, even into the Hoofs, and in their Progress knit every Joynt together. Besides what I have named, there are two main Sinews coming through two Holes to the flat Bones of the Hips or Huckle; and from thence, being divided into many Branches, pass down into the hinder Legs, even into the Coffin of the Hoofs, binding all the joynts firmly together. Now, above all these, there is one main Sinew of Ligament proceeding from the setting-on of the Horse's Neck, which passes along the Chine, and is one intire Sinew of near three Inches broad, being in its nature flat, and not having any Branch proceeding from it: And this is the great strenthener of the Back and Neck of any Horse, and holdeth the Shoulder-blade firm; so that in all, it's reckoned, a Horse hath no less than thirty principal Sinews, whence a great number of other Sinews do proceed of lesser note. But these which I have named are so useful in supporting the frame of the Body, that if any one of them are afflicted, the Horse will be much afflicted, *&c.*

And now the next thing material to be known of this kind is, the Number of Bones, and their Situation; which, according to the best Account that has been given, take as followeth:

The full number of Bones found in the Body of a Horse are a hundred and seventy, *viz.* The upper part of the Head are computed two Bones; from the Forehead to the Nose are two more; the nether Jaws are computed two Bones; the four Teeth are found to be twelve, and the Tushes four, and the Grinders twenty four; the Bones in the Neck, from the Spade-bone to the Huckle-bone eight; from the Hucke-bones to the end of the Tail are accounted seven: Next, or at least next to be considered, is a broad Bone, in which are found twelve Seams or Joynts: Then the two Spade-bones, and after them the Forcales or Canale-bones; and subsequent to these, those that are commonly called the Marrow-bones, and from thence to the Knees two called the Thigh bones, next to them two others, proceeding from thence to the Pasterns, called the Shank-bones, and so downward into the Hoof; other Bones, called the Supporters, in all sixteen, though some of them are very small. In the Breast there is a great Bone, whereunto thirty six Ribs, great and small, are appendances; and to the Colume behind are fixed two Bones, and from the Molars to the Joynts you will find other two Bones, and two more towards the bending of the Ribs; from the bending of the Ribs, unto the Ribs, are two Bones, though very small in comparison of those we account Capital ones; and from the Legs to the two Focils of the Legs, are to be found other two little Bones. And moreover observe, that from the Pasterns to the Hoof, and in the Hoof are sixteen little Bones: all which are ruled and governed by Ligaments, and tends on to the best advantage or complacency, in the frame of Nature, and are necessary to be known as to the frame and station wherein they reside, or are situate.

Chapter XX

Of Blood-letting in general, and how to know when and where it is convenient to let Blood, for preventing of Sickness, or recovery of Health.

To let a Horse Blood too often, decays Nature, and much weakens the Horse; when on the other side, too long to defer it, gives corrupt Blood opportunity to create Diseases that infect and afflict him; if it be done for Health's sake, without any sign of apparent danger, then is twice in a Year sufficient: The times best approved of for this office, are the end of *December*, and the beginning of *May*. Now some are of opinion, that there is no need of letting a Stone-horse Blood, if he be used to cover Mares, unless some disease be apparent; and the reason they give is, that his spending does exhaust his Blood sufficiently; but such a reason is very weak, considering the Blood of which the Seed is made, by the working and operation of the Genital Parts, is the most spirited, and pure of all other; and that which ought to be taken away by Blood-letting, is commonly the most gross and offensive, so that it will never turn to Seed, but corrupt by continuing in the Body, and create Diseases; and therefore ought it to be drawn off, that the Veins being emptied, may fill with better Blood; nay, Blood-letting is coveted naturally by some Horses; for the Horses of *Poland* often let themselves Blood by often rubbing against a sharp Post or Rock: and so does the River *Horse* in *Nilus*, from whose Example 'tis held that Phlebotomy or Blood-letting came in use. Others there are that will not have Geldings let Blood; and the reason they give is, that his Body, through the loss of his Genetors, being infeebled, wants the Blood to support and corroborate it, and that consumes faster in him than in a Stone-horse: but this bears no weight, for the Blood will increase according to the constitution of the Horse and his feeding, and according to the heat or coldness of the Country the Horse is bred in, the less or more Blood is found in him.

The next material matter to be regarded, is, the exact time when it ought to be performed; and most hold it best in the Morning, when the Horse is fasting, an hour after he wakes or rouses up: and then again have regard to the Moon, it being in the increase, if possible, but by no means in the Sign where the Vein is situate that you open, lest thereby your Horse be weakned, for then the Vital Spirits will issue with the Blood. And another regard is to be had to his Age; for if he be very old, you must not bleed him, unless upon some emergent occasion, but suffer his Blood to nourish him, that his flesh may not wain or decrease; and in all these cases you must have regard to a Horse's Constitution, for some are more capable of losing three pound of Blood, than others are two or one; or in case of a Disease according as more or less is required, as the Blood is more or less infected; and in such a case no time nor season is to be regarded, but as necessity requires; for many times Blood-letting in such cases is a means to save the Horse, as deferring of it is to destroying of him: and in these or the like cases it ought to be known in what Diseases it is requisite, and that it may in the general be discovered by these Signs, *viz.*

If the Horse's Eye look red, and his Veins rise and swell beyond the ordinary bounds, then is he oppressed with too much Blood, or at least that which is not good. If you perceive by his itching, scrubbing himself, and uneasiness, that there is a salt fiery Humour contracted in the Veins, then let him Blood, which is more immediately known by the extraordinary itching of the Mane and Tail; by rubbing of which, the Hair frequently sheds, and the Skin peels off. If his Urine be red and high-coloured, and his Dung very hot, black and hard; if on his Back there appear little Bubbles or Inflammations red and angry; or if his Meat be not well digested, then they denote the Horse to abound with Blood, and stand in need of having it drawn off to a degree, that his health may be preserved or restored. If the Whites of his Eyes are tainted with yellowness, or the like happen to his upper or nether Lip, it is requisite to bleed him; for upon the happening of any of these Signs, the Blood is disordered, super-abundant or corrupted, and some fatal sickness is threatned; to prevent, or at least mitigate which, there is no better way than Bleeding. And in the manner of letting Blood, when you have corded your Horse conveniently, that the Veins may appear full and fair, you may strike your Fleam in the Neck-veins four Inches on this side of the setting-on of the Heat; but the other Veins being small, you must open them with a Lancet, for fear of cutting them through, and so consequently injure the Nerves that support them. And thus much for Blood-letting, and the Signs of Bleeding, *&c.* From which I shall proceed to what relates to Cures of all kinds, whether internal Diseases, or external Sorrances, after the best and easiest manner and method, according to the approved Rules and Directions of the ablest Farriers and Horsleeches that are, and have been rendred famous for their Skill, not only in this, but in divers other Nations; and withal let you plainly understand the cause of the most chronical Diseases as well as the means to Cure them, rendring the method and manner easie to the Practitioner, and advantageous to those that trade in Horses, and keep them for their profit or pleasure. But before I absolutely enter upon them, there are a few things necessary to be considered, and especially five, *viz.*

1. To inform one's self to what Grief or Distemper a Horse is inclinable.
2. The Cause from whence it proceeds, whether external or internal.
3. The Tokens and Symptoms by which the Distemper is known.
4. By what means the Causes accrue that create the Distemper.
5. How to apply Remedies fitting to the Distempers and Disorders.

And these are so absolutely necessary, that without such Knowledge a perfect Cure cannot be wrought by any Farrier.

And now as for the Diseases and Grievances incident to a Horse, they are not a few; those that reckon the least, consent to sixty of different natures: and, to be plain, there are many more which may be called Dependants on these, or indeed some of them no less dangerous.

Chapter XXI

A Description of Diseases, Grievances, or Sorrances, incident to Horses, &c. to foresee them by sundry Signs and Tokens; and know whence they arise; with the Ways and Methods of Preventing, Redressing and Curing them, by approved Rules and Remedies.

MANY ARE THE TERMS AND NAMES given to Grievances and Sorrances, and these frequently take their denomination from the place of their situation, relating to the parts of the Body, and chiefly arise or proceed from the evil state or habit of the Body and defect of the Members, especially when there happens a loosning and division in the Unity, or a Dissonance in the Temperature: and so we say when any division or loosning happens in the Bone, it is a Fracture; if it happens in the Veins, it is a Rupture; and in the Flesh, a Wound or Ulcer; in the Sinews it is termed the Cramp or Convulsion; and in the Skin an Excoriation. And now observe, that in Cauterizing or Burning, which may be done two manner of ways, *viz.* with hot Oils, or Water, or with a hot Iron; as also in making Incision, or the like, that they ever be done with caution above or beneath the Vein or Sinew, and never upon them, lest by the discommodity that may arise, the Horse be rather damaged than furthered: And again, observe never to apply to Veins or Sinews any extraordinary Corrosives; and these we reckon so, are *Arsnick, Sublimate Mercury, Resalgar, Oil of Vitriol, Tartar,* &c.

In case of Cauterizing, ever observe rather than to launce with a hot Iron than a cold one; and rather cauterize than cut, because by the heat the humours are driven back, which would otherwise render the Sore or Grief of long continuance; and by the product of a Sore in the original, you may know the Consitution of a Horse; for if he be a Sanguine, a whitish watery Excressence will flow of an indifferent thickness; if Cholerick, a thin salt one, inclining to greenish; if Flegmatick, a kind of gellied-water; and if Melancholick, then a gross blackish moisture, attended by a dry Scab or Scurf.

In case of Swelling or Tumours, it will be necessary with heed to mark the place they begin in, as also their progress and ending, that so the part of the Body afflicted with bad Humours may have Applications sutable to remove them.

Now if it so happen that the Swellings or Tumours be not upon any Master-vein or Sinew, or too near the vital Parts, then may Repercussive Medicines be used: but if so, then those that are mollifying, to raise it gently by degrees, that so being ripened it may break of itself; or if necessity require it, be lanced, that the Corruption passing away such things may be applied, that with a lenitive softness may heal the grief.

In such case the swelling be hard, it is generally Corrosive, and must have such Medicaments applied: but if soft, those of a more supple Nature will prevail, though many happen to be of a long continuance. And now to know whether a Swelling has been of a long continuance, press it with your finger; and if the dint continue after you have taken your finger away, then is it old, and of a long standing; but if the flesh suddenly returns, then is it newly taken. If a Swelling or Sore be broken, and moderately matter, then it signifies a good Constitution, and that it mends apace; but if the Putrefaction greatly increase, then it denotes a mass of bad Humours there contracted, which must by degrees be brought away; and many times in this or the like cases is Cauterizing actually or potentially used, that is, with hot Irons, or with Liquids, &c. above or beneath the Sore, to divert or keep back the Humour.

Chapter XXII

Excellent Receipts for the Cure of Diseases, &c. in Horses or Mares, according to the best Experience of Skilful Practitioners.

As the Causes of Sickness are many, so are the Cures or Medicines wherewith they are redressed, and the Signs that fore-run and attend them, which I have given in the Generals and Particulars. Yet seeing something of this kind may remain untouched, I shall take them in the way as I proceed.

The Glaunders, from what it proceeds, and how to Cure it.

The Glaunders is a Distemper often gotten by over-riding, and suddain cooling upon Heats; eating too much raw food at unseasonable times, or continuing in wet Moorish ground, proceeding from a flegmatick Constitution: and this you may perceive before it take too firm a possession by a white matter issuing from the Horse's Nostrils, and his unseasonable snorting: And this Disease having its seat properly in the Head, Take the green leaves of a Box-tree, an Ounce of Aninseeds, and the like quantity of Licorish-powder: steep them in Ale or new milk, to the quantity of a Quart; adding Treacle four Ounces, and the like quantity of Olive-oil: heat them well over a gentle Fire, and with a Horn give the Horse the liquid part to drink in the

Morning fasting, as hot as may be well endured; then give him a moderate Course, and bringing him home, let him have a warm Marsh; cloath him up, and leave him in a warm Stable for the space of two Hours before you feed him.

The Quinzey in a Horse, &c. what it is, and how to Cure it.

The Quinzey is a Distemper, occasioned by flegmatick Humours, settling in the Neck and Throat, so the passage of the Breath is in a manner stopped, and the Beast obliged to breath with a more than ordinary Pain and Labour: To remove them then, and redress the Grievance, after having bled him in the Neck-vein,

Take Marshmallows, Groundsel, Camomil and Hart's Tongue; bray them with a like quantity of Smallage, and fry them with hogs-grease; so being very hot, apply them to the place which ought to be under the Chaul, and it will mollifie the swelling, and by degrees remove the obstruction of the passage: Then take Rorch-allom, Honey and white Dog's Turd; dissolve them with brown Sugar-candy in a quart of Milk; give it him hot, and so continue to do for a Week, Morning and Evening; renewing likewise the Poltis once a Day.

A Horse's Bleeding at the Nose, how to stay or prevent it.

This happens, especially amongst young Horses, through the abundance of Blood, that through the free passage of the large Veins, ascends into the Head; and passing to the thin Veins within the Nostrils, either by its violent motion forces them, or by its corrosive quality eats them in sunder, or it may accidentally happen by a stroak or violent straining: To remedy which,

Take the Juyce of Nettles, mixed with Loaf-sugar, and squirt it up the Horse's Nostrils, using at convenient times to burn under his Nose Storax, Frankincense, or Linnen dipped in *Aqua-vitæ* in a Chasing-dish, the fume of which will oblige the Blood to retreat; or instead of Nettle-juyce, you may use that of Garlick, blowing up after it Powder of dried Rhubarb.

For Pains in the Teeth or Jaws.

Pains in the Teeth or Jaws of a Horse are created by Wind or cold Rheums that often afflict the Marrow or Sinews which support them;

and are often known by the Horse's driveling, holding open his mouth, and defect in feeding: And in this case,

Take a good handful of the Herb Bettony, seethe it in a quart of White-wine vinegar till half be consumed, then add an ounce of the Power of Elecampane; wash, with the liquid part, his Teeth and Gums, anointing his Jaws with Oyntment of Tobacco, lancing the Gums, and opening the Temple-vein on the side where the Pain is most predominant.

The Canker in the Nose, what it is, and the Remedy.

The Canker in the Nose proceeds from a virulent Humour conracted there, occasioned by Inflammation: To cure this, or, indeed, one in any part of the Body,

Take of White-wine-vinegar a quart, Roach-allum two pounds, a pint of the Juyce of Plantane, and as much of that of Rue, with four ounces of Honey; boil them to the consumption of a third part, and wash the grieved part therewith, as hot as the Horse will endure it, morning and evening, and the Canker within a fortnight will decay.

A Remedy for the Chollick, Belly-ach, or Belly-binding.

The two first of these proceed from cold, raw, slimy Humours, settled in the Bowels, there occasioning a fretting or shrinking up of the Bowels, or by a contraction of Windiness, which stretcheth them out; and the latter by excessive heat or bad digestion: Now to remedy these,

Take four ounces of Dill-seed, or the Herb itself, if it may be gotten; and also two handfuls of May-weed: boil these in a Gallon of small Beer, with a pound of brown Sugar, and give the Horse a quart at a time, each morning fasting, for the space of four days, suffering it to be lukewarm.

The Lasks, or Bloody-flux, and its Remedy.

This Distemper is an unnatural looseness of the Body, contracted either by raw feeding, or a fretting of the Bowels, caused by cholerick or fiery Humours, and a superfluity of bad Blood, which not being stayed in time, will, for want of other Excrements, cause the Horse to void Blood: To Cure this,

Take a good handful of the Herb called Shepherd's Purse, and two Ounces of the Seed of Woodroof; stamp them, and boiling them in three Pints of Ale till it come to a Quart, strain out the liquid part, and give it him hot in a Drenching-horn, continuing to do so for three or four Mornings successively.

The Botts, what they are, and their Cure.

The Botts are a short thick Worm, like a Maggot, having black Heads, and are engendred through the Corruption of heat and moisture in the Maw or Bowels of a Horse, where they knaw and afflict him in a grievous manner, and may be discerned by his lifting up his Feet to strike at his Belly, and the small Stomach he has to feed: The way to destroy them is thus,

Take Rue, Savin, Nightshade, the Seeds of Ameos, each two Ounces; bruise them well, and with Honey and Roach-allom make them into little Balls, and buttering them over, suffer him to swallow two of them in the Morning fasting, and about an Hour after give him of Salad-oil and *Aqua-vitæ*, each a quarter of a Pint, very hot; and after that let him stand another Hour before you give him any Provender; and this Rule observe for a Week together.

For the Shoulder-strain, a good Remedy.

Having tied up the lame Leg, drive your Horse till, by his striving to go on three Legs, the Plate vein appear; then let down his Leg, and Bleed him on that side, having in the readiness a Bason with Salt in it to receive the Blood, stirring it to keep it from clotting; then having likewise an Ounce of the Oil of Turpentine, put it into half a Pint of Beer or Ale, warm them well; and when he has bled a Quart or more, as you see occasion, nip the Orifice with a piece of Lead made in form for that purpose, and chafe the Place grieved with the Oil and Beer, and afterward with the Blood and Salt; then putting him into a warm Stable, draw the two Fore-legs into an evenness with a Lift or Girt, binding them strait, and so permitting him to stand; or, if you think convenient, to keep his Feet the firmer and evener, wedg them between the Shooe and the Hoof with a flat stick, continuing to anoint and bathe him as often as occasion requires, unbinding his Legs, and leading him each Morning till you find the Lameness decrease: And thus you may do for a Wrench or Distortion of the Shoulder.

Broken wind, what it is, and to remedy it, if not past Cure.

The Breaking of the Wind is occasioned by excessive Riding or Straining, or by bad Usage after much Labour, or excessive Rheums falling upon the Lungs, and hindring them in the performance of their Office, to that degree, that the closures of the Windpipe shrink up, or are restrained from their former pliableness: And to remedy this, if not too far gone,

Take the Soil or Dung of a Boar, or Barrow pig, dried, that it may be reduced to Powder, and of Anniseeds an equal quantity; boil them in Milk or Whey, and give him a Pint of it hot to drink every third Morning, causing him to Exercise moderately thereupon: or it would be better if you could get the Dung of a Hedge-hog, usually called an Urching; the Dose to the Pint of Milk is two Spoonfulls of each of the Ingredients.

A Horse Burnt by a Mare, how to Cure.

This Grief is to be observed by the Mattering of the Yard, *&c.* and the dullness of the Countenance of the Horse. Which perceived, to remedy the Grievance,

Boil a quarter of a Pound of Roach-allom in a Pint of White-wine; and being reduced to a thinness, squirt part of it, Blood-warm, up the Horse's Yard with a Syringe; and by doing so once a Day, for a Week together, the Cause will cease, and the Effects be no more.

For a dry Cough, Cold, Pursiveness, Broken-wind, or Shortness of Breath.

These Distempers being occasioned by cold gross Humours, rawness of Diet, over-much Labour, unseasonable gross Feeding, and unwholesome Airs which infect the Lungs:

Take an Ounce of Rue-seeds, Tarr two Ounces, and as much fresh Butter; mix them well together with the Powder of Licorish, Anniseeds, Nutmeg, and brown Sugar candy, and make them into three Balls in equal Portions; and when your Horse is warm with motion or riding, having put into each Ball three or four Cloves of Garlick, oblige your Horse to swallow them when he is fasting; and bringing him home, let him stand without Meat for the space of an Hour in a warm Stable. If the former fail, as rarely it does, Take of Bacon-lard a piece to the breadth of four Fingers, and every way as thick as long, if you can get it; but as for the thickness, two Fingers may serve; then stop it with Cloves and Garlick, dusting it over with Powder of Licorish, Anniseeds, Sugar-candy and Flower of Brimstone, and cause him, in two long slices, to swallow it fasting, and ride him thereupon, that he cast it not up: and do so every Morning for a Week together, giving him after it a Glass of Malage.

Now to make a Horse in this case Swallow, though against his will, Draw forth his Tongue as far as is convenient, and put the Potion down his Throat beyond the roughness; and then suddenly letting go his Tongue, he will swallow it done without tasting or scenting; and in so doing, you must ever draw up his Head to the Rack, that it may descend the better.

To restore decayed or putrified Lungs.

The sign of the Lungs being in this Disorder, is to be known by a faint Cough, and the casting of putfried matter out of his Mouth like small pieces of red Flesh, eating his Provender with greediness. To redress this,

Take the Juyce of Purslain, or for want of it, that of Housleek, half a Pint, Steel-dust two Ounces, Oil of Roses four Ounces, of Tragcauthium one Ounce, add to them a Quart of Goat's Milk, and give it him hot at three times, keeping him for a while after fasting and in motion: And this you may use till you find his Breath become sweet, and the Couch allayed; the which, if it be not too far gone, will be in a Week's time. And the better to refresh the vital Parts, you may burn under his Nose *Storax, Galbanum* and *Myrrh.*

A dry Consumption, its Remedy, &c.

This Disease is occasioned by sharp corroding Humours descending from the Head, and falling upon the Lungs, by which they are many times ulcerated, and by their bad effects cause a macerating or wasting of the Body, yet sendeth forth no Corruption at the Nose, because the moisture is consumed by the Heat: To remedy which,

Take a Pint of the Juyce of Comsory, half a Pint of Oil of Roses, the Juyce of four Lemons, and an ounce of the Juyce of Rue; let them simper over a gentle Fire, and add the Powder of Round bithwort roots two Ounces, and an Ounce of that of Rhubarb, and give him these in two equal potions Morning and Evening.

The Breast-pain, from whence it proceeds, and how to Cure it.

This Disorder of the Body proceedeth from a superfluity of Blood, which presses the Heart, and gross indigested Humours that make the like unnatural invasion upon the Liver; and the signs to know when a Horse is afflicted, are a stiff staggering and keeping together the fore-legs as it were, and but weakly proceeding in his pace, *&c.* his gate distorted and uneven, stifly, for the most part, holding up his Head and Neck, as not being well able to reach the Ground; and moreover, you will observe him to groan and strain in his eating and drinking.

To Cure this, which frequently ends in Death, if not timely regarded, let him Blood in both the Breast veins; and when he has bled sufficiently, two quarts at least, chase his Breast and Fore-body with Oil of Peter, that the Blood may be drawn into the empty Veins, and so ease the vital Parts of their Oppression, and give him a Pint of warm White-wine, with two Ounces of *Diapente*: or if the Pain afterward continue, which is very rare, you may towel him.

Heart sickness, or Antecor; *whence it proceeds, and how to remedy it.*

This Distemper being an Enemy to the Heart, seems from thence to take its denomination, and is caused by a superfluity of the Blood in Horses that are fed high, and are put to little or no Labour; which Blood, for want of Motion, being corrupt, infects the Heart, and renders it sickish and languishing.

The Signs are a small swelling, rising at the bottom of the Breast, increasing upward to the top of the Neck, whither if it arrive before it be prevented, Death frequently issues; it is known also by his hanging his Head, loathing his Food, and groaning through the Oppression of Pain. These things being observed,

Let your Horse blood on either Plate-vein, or in the Neck, if the Swelling be risen high; and having bled him freely, Take of the best Malmsey a Quart, add it to two Ounces of Sugar, and an Ounce of beaten Cinamon; give it him to drink Blood-warm, and it will revive and cherish the Heart by dispelling the evil Vapours from the seats of Life; and after that wisp Him well over his Cloth, and let him rest, giving him the next Morning the like Dose, and riding him gently.

Foundring in the Body, or Surfeiting, how occasioned, together with the Remedy.

This Disease, according to the Skilful, is occasioned by a contraction of Molten-grease and evil Humours which oppress the frame of the Body, and is taken by the bad management of those that use the Horse indiscreetly; watering him when hot, or letting him upon a heat suddainly cool in a bad Air, or moist Place; by which Humours, together with the Molten-grease, have opportunity to contract themselves into one Mass, to that degree of Consolidation, that Nature without extraordinary Helps, is not capable of dispersing or dispelling them.

The Signs demonstrating this Distemper or Grievance, are the flaring of the Hair, hanging of the Head, an unusual Cough, staggering, belching, the clinging up of the Belly, and rising of the Back.

To remedy this Distemper, Take a handful of Mallows, as much Smallage, Camomil and Groundsel, an Ounce of Alloes, two Quarts of new Milk, and half a Pound of brown Sugar; boil them together, and strain out the liquid part, giving it to the Horse Clyster-ways; and when it has caused him to empty sufficiently, take a Quart of Malmsey, or, for want of it, Canary, of Licorish, Anniseeds and Cinamon beaten to Powder, each half an Ounce; put them, with two Ounces of beaten Sugar candy, into the Wine, and give it the Horse warm, keeping him afterward in Motion, tho' in the Stable, for the space of two Hours, well cloathed and littered; after which give him two Quarts of Oats, and a Quart of Splent-beans, well sifted and sprinkled with Beer, *&c.*

The Greedy Worm, or Hungry Evil; what it is, and how to be remedied or prevented.

This Distemper, or rather Defect in a Horse, has deceived many; for whereas they take it for a good sign to see a Horse feed hastily, and be voracious, yet is it the cause of Sickness, and decay in the end, if not prevented; for the cause a Horse is so hasty and large a Feeder, is either by reason he has been a long time debarred from Meat, and so the Veins being open and empty, crave nutriment; or from too excessive a Heat in the Stomach, that consumes the moisture and nutriment faster than it can be reasonably expected to digest: The first of these, if not prevented, creates Diseases, and other Disorders, by the Veins, drawing the crude Digestion not perfected, and filling their Cavities with a gross watry Humour, rather than good Blood; and the latter by feeding the Heat (which otherwise would decay) that consumes the radical Moisture.

The help for this is, first, feeding the Horse by degrees, suffering a regular Digestion; and the latter by giving him cooling things to allay the Heat, and moderate the Appetite, are the best things to reduce him to a due temperature, and managed Diet, as has been said; but more particularly, give him a Quart of Cream, a Pint of White-wine boiled with a handful of Wood-sorrel, and the like quantity of Scabeous or Mugwort, the liquid part only: Let him take it cold, and rest upon it, and it will close the Veins to that degree, that the Digestion may be made perfect, as also it will allay the Heat.

Yellow and Black Jaundice in a Horse, the Cause and Cure, &c.

The Yellow Jaundice generally arises from the abundance of cholerick Humours contracted, which occasion the overflowing of the Gall, and are great Oppressors of the Body, and Obstructors of Health; and the signs are the yellowness of the Mouth-skin, insides of the Lips and Eyes. There is another kind of this Disease that proceeds from Melancholy, and these are called the Black Jaundice, and have their original from obstructions in the Liver-vein, which passeth to the Spleen, and consequently hinders the Spleen from doing its proper Office, by receiving only corrupted Blood from the Liver; and so is obliged, by reason of its being surcharged, to cast it back into the Veins; and this latter in case of Death, as indeed being most dangerous, mastereth the former; but a timely regard may remove them, as thus:

Take, after you have blooded your Horse in the third Barr of the Mouth, an Ounce of Turmerick, and half an Ounce of Saffron, four or five Cloves, and six Spoonfuls of strong Vinegar, Long-pepper and Licorish, beaten fine, of each an Ounce, with the like quantity of Burdock-roots; boil them in two Quarts of Ale till a third, or at least a fourth part be consumed, and give it him to drink very hot; and in so doing you will find the bad Humours disperse, and by degrees losing their force by the operation of Nature, and the conquering quality of the Medicament; but if it be too long delay'd, it many times runs beyond the help of Art.

Costiveness, from whence it proceeds, and its Remedy.

This Disorder is a hardening of the Excrements in the Body, so that without great Pain the Horse cannot evacuate or void his Dung; and this is often occasioned by the excess of Provender, insomuch that Nature forces it into the Bowels before it be well digested in the Stomach: or again, it happens by feeding altogether upon dry Meats; the which, though wholesome and nourishing, contract notwithstanding the Excrements by the extraordinary Heat they occasion; and it may likewise happen by excessive Fasting. To remedy which, you may give him a Clyster made in this manner:

Take a handful of Mashmallows, decoct them in Spring-water, not exceeding a Quart; add to these half a Pint of Salad oil, and six Ounces of fresh Butter, of *Benedict a Laxativa* an Ounce, and force them warm up his Fundament, holding or tying close his Tail, by bringing it with a Cord between his Legs, obliging him to keep it in for the space of an Hour: and the better to make it work, give him a warm Marsh, and as soon as he has discharged it, give him in a Drenching-horn a Potion made after this manner:

Take two Ounces of Castle soap, dissolve it in a Pint of warm White-wine, and with it a quarter of a Pint of Linseed-oil, sweetning them all with Sugar-candy, and give it him as hot as he can drink it.

The Cramp, or Convulsion in the Nerves, or Sinews, how occasioned, together with the Remedy, &c.

These generally proceed from some bruise, wound or other hurt on the Nerves and Sinews, or excessive straining, especially when the Horse after a great heat by riding or servile labour suddenly cool'd.

This Grievance is known by the trembling of the Joynts, Nerves or Veins, or by their contracting to such a stubbornness or stiffness, that, for a time, neither the Horse, nor those that attempt to help him are capable of bending them.

To Cure this, Take Camomil, Primrose-leaves, the Roots of Crowfeet and Cowslips, with the Branches of Fennel, Rosemary and

Pimpernel; boil them in Running-water, and having pressed out the liquid part, bathe the Place grieved with it exceeding hot, binding on the Herbs Poltis-ways with coarse Linnen, or Bands made of Straw or Hay, and keeping him in a warm Stable, with good Provender, his Limbs will be restored, and rendred as before; and the better to hearten him, let him receive for a Morning or two, the Yolk of an Egg in a Glass of Canary.

The Mourning of the Chine, its Cause, and the Means to Cure it.

This Grievance is caused by suddainly cooling upon excessive Heats, standing in damp or wet Places, or eating such things as turn to raw Humours; which falling upon the Liver and Lungs, frequently inflame or putrifie them, so that they occasion the Horse by defect of their office to fall down suddainly and die: Therefore when you by any trembling or dullness suspect this Grievance, let your Horse blood, and having chafed him well,

Take Olive-oil and Verjuyce, of each two Ounces, the Juyce of Sellendine, and Powder of Elecampane-roots, of each an Ounce; warm them a little, and tying his Head up to the Rack, pour them into his Nostrils, stopping them close after it, that he may be forced to sneeze and strain to cast it out; after which having an Ounce of the Powder of Rhubarb heated in a Pint of Canary, give it him in a Drenching-horn as hot as he can well endure it; and so use him each Morning for a Week together, and the bad Humours will be worked off.

Frenzy or Madness, its Cause, with the Means to remedy it according to the experienced way.

This Disease is very dangerous, and often terminates in death, and is occasioned by hot and fiery humours unseasonably mixing with the blood, which by its ascending inflammation afflicts the brain, that principal seat of life: and this is known by the staring of the Horse, the distorting of his eyes, hanging of his ears, staggering and giddiness, his often crying and forsaking his meat; and if it be wrought to a height, his often beating himself against the posts, manger, and other places he can conveniently come at; biting, stamping, and flying about; with many the like disorders.

To Remedy this, speedily let him blood in the temple-veins; and if he bleed not freely there, strike him in the neck-veins; when having bled sufficiently, Take the Roots of Gourds or Wild-cucumbers, black Hellebore, Rue and Mint, with *Virgo Pastoris*, of each a handful, boil them in Beer or fair running Water, and give him the liquid part very warm; and doing so three or four times it will purifie and purge the Blood: but if you suppose it too weak for the Horse's constitution, you may dissolve in it an ounce of well-washed Allows. And observe in this case, above all things, to keep him warm.

The Falling-evil, its Cause and Cure.

This Distemper is caused by a vapour that oppresses or annoys the vital parts, rendring the frame of the body for a time senseless, and altogether unable to distinguish what befalls it, and has its original from an evil habit of body; and its approach is frequently known by the coldness of the nose and gristles thereabout.

The speedy, at least the best Remedy, is to let him blood on both the neck veins, in the morning when he is fasting, and then prepare a dose of Powder of dried Berries of Mistletoe, and the Powder of the Hart's horn, each an ounce; of the Oil of Nutmeg and Pepper, each a dram: compound them in half a pint of Canary, and give it the Horse, when you perceive the grief to be coming on him, as warm as may be.

The Sleeping evil, what it is, and the way to remedy it.

This is a Distemper, frequently occasioned by the over-moistness of the Brain, or rather a watry coldness contracted within the Cells, which chills and numbs the Brain, whereby the Horse becomes dozed, heavy and stupid, ever desirous to sleep, yet still troubled with restless Dreams and Disorders, and owes its original to moist feeding in Marshey-grounds, whereby abundance of phlegmatick and watry Humomurs have been contracted: And in this case likewise letting Blood in both the Neck-veins in much available. But further to perfect the Cure,

Take Camomil and Mother wort, of each a like quantity, boil them in a Gallon of Running-water, with a pound of Treacle, and a handful of Bay-leaves, and give the Horse a pint each Morning fasting, as hot as he can endure it, keeping him very warm, and fasting for the space of an Hour after; and then of Malt or scalded Bran make him a warm Marsh.

The Horse-Pestilence, and its Cure.

The Pestilence in Horses is either contracted by bad feeding which occasions a Corruption or Inflammation of the Blood whereby the Heart is afflicted, or it happens by being in foggy and infectious Airs, or catched by Contagion: And in any of these cases,

Take Lavender of a handful, the like quantity of Rue and Wormwood; as also of Walnut-tree-leaves, and an Ounce of Alloes; boil them in a Quart of Water, or three Pints of Milk, till half be consumed; then add half a Pound of fresh Butter, or rather, if you can get it, half a Pint of the sweetest Olive-oyl, and straining out the liquid part, give it the Horse fasting in the Morning Blood-warm, repeating a fresh Dose every other Morning, for the space of a Week.

For Chest-foundring, the Remedy.

To know whether your Horse be Chest-foundered, or not, observe him standing; and if then he do as it were stand drawn up, or crimpling with his Body, or stragling and covet much to lie down, running sometimes backward in his going, then it is apparent he is afflicted with this Grievance: To Cure which,

Take Oyl of Peter half an ounce, mix it with an ounce of the Oyl of Camomil; and so proportionably a greater quantity, as you see occasion, and bathe the Breast with a hot Woolen-cloth; and when you have in that manner chafed it as well as you can, run a hot Iron over it to make it sink into the Skin: do this twice or thrice, and give the Horse a quarter of a pint of Salad-oyl, and the like quantity of *Aqua-vitae,* warmed and well mixed together over a gentle fire.

For an Obstruction in the Bladder, or Windiness in the Bowels, use this approved Medicine.

Of Cake or Castle soap take twelve ounces, scrape it so that it may be rendred very small, adding two ounces of *Dialthea*; incorporate them well, and make them up into balls as big as Pigeons Eggs; and when you find your Horse afflicted, as aforesaid, dissolve one of them in a pint of Ale or Beer, and give it him scalding-hot, or so hot as he can take it without danger, and it will force a passage for the Urine without much difficulty. This is also good for the Stone, or Gravel in the Kindneys.

The Pole-Evil, how to know and Cure.

The Pole-Evil is known by its growing bigger than ordinary on the top of the Head; where, if you find it large, take a hot Iron, and Sear it in a Circle, after the form of the Figure, till the Skin become as it were of a yellowish colour; then with a sharp Iron make holes in it, one large one in the middle, small ones circling it within the first Circle; the form of the Iron, which must not penetrate above half an Inch, you have in the Margin. The holes made as directed,

Take a piece of yellow Arsnick, to the bigness of a Pea, and divide that, or a somewhat larger quantity, that apart may be applied to every hole made, as aforesaid, and cover it over with black Soad, then with Hog's Lard and Verde-grease, anoint the rest of the place seared, and cover it all over with a Cloth dipped in Oyl of Turpentine; and so by the corrosive nature of the Arsnick, the contacted swelling will be so loosned, that with a little cutting or drawing off the bottom, the Core or Cause of the Grievance may be drawn out or taken away, and this frequently, especially according to the Constitution of the Horse, may be attempted in a week or ten days after the Application; and having washed the wound with Plantane-water, wherein a small quantity of Allom has been dissolved; anoint it with Oyl of Roses, or Ointment of Tobacco, and cover it up close from the Air, anointing it once a day, till the flesh fill up the hollowness; and if proud flesh appear, notwithstanding scald it with Salt and Butter.

The Festula, how to discover and cure.

A Festula is the contraction and settlement of bad Humours, or Infection into one place, occasioning an ulcerous Tumour, and is best suppressed by Cauterizing in circling it round, to prevent its farther spreading, and likewise to deny the Humours that feed it access; and when you have with a hot-Iron circled it as the former, prick it full

of holes with a three square sharp Instrument; the Figure of which, and of the Circle, take notice of in the Margin, and so use it in all respects, as that of the Pole-evil, if you find it very corrupt; but if it appear shallow, mitigate the Corrosives to half the quantity, and search it in a shorter time, letting out the Corruption, if it will come forth by applying Lenitives, &c. And when you find it begin to heal, anoint it first with Oyl of Camomil, and after that take the effects of the fire quite away, with Oyntment of Marshmallows, beaten, with the White of an Egg, or *Spermacui*.

Hard Kernels under the Throat, how to remove them.

Take half a pint of Brandy or *Aqua vitæ*, put into it a quarter of a pound of common Soap; boil them till they become thick as a Plaster, and apply it plaster-wise to the place grieved; and if no store of corruption, or a contraction of evil attend those Kernels, then will it sink them, so that they will not be offensive; and if there be humours, it will break and disperse them.

For the Navell-gall, the Remedy.

Take an indifferent fine rag, dip it in Brandy and Sallad-oil, well incorporated over a gentle fire; bathe and supple well the place grieved; and to make it penetrate the better in often so doing the cure will be wrought.

For a Blow, Bruise, or the like Misfortune that causes a Swelling or Tumour, the Remedy.

If the swelling be about the head, let the Horse blood in the neck-vein, on that side the misfortune befell: which done, to prevent the Farcy, and the like,

Take of Anniseeds, Rue, Turmerick and Red-sage, each about an ounce; shread them out into a quart of Beer or Ale, and suffering them to infuse therein for the space of a night, press out the next morning the liquid part very hard, and give it him cold to drink, suffering him to fast after it for the space of four hours; then having in readiness a

Charge made of *Aqua vitæ* and Soap, spread it upon a Leather, so much as will cover the swelling, and your expectation will be answered.

The Scratches, their Remedy.

The Scratches are a troublesome Sorrance found upon the hinder heels of a Horse, on the pasterns and somewhat above, and are caused by the breaking out of evil humours settling there: To cure which,

Take of Hen's dung and Black-soap, each two ounces; incorporate them with Hog's grease, or Neat's foot-oil, over a gentle fire, till they become an Ointment: then, having dried and rubbed the Horse's heels, anoint them with it, and bind on the Ointment, or swathe the legs with a warm cloth, not suffering the Horse to come into the water. And if this prove not a sufficient Remedy at several times using, try this more powerful:

Take Beef-broth and bathe his legs well therewith overnight, and rubbing them clean the next morning, take two ounces of Deer's suet, the like quantity of Speck-oil, and an ounce of Verdegrease; beat them well together in half a pint of Train-oil; put them into an earthen pot on a gentle fire, and stir them well together, and anoint the place grieved, chafing it in with a hot Cloth, keeping him out of the Water and dirty Ways.

For Foot-foundring, a Remedy.

Having found by the lameness or cripling of your Horse, that he is foundred, which mostly happens by unseasonably travelling in dirty Ways, and not being well regarded upon his setting up; bleed him a little in the Thigh, or if you can conveniently, in the Fetlock vein, and set on his Shooe hollow, that Wool or Cotton may be thrust between as occasion requires it; then

Take *Venice*-Turpentine, and spread it upon a lock or wad thereof, putting it with a flat stick between the Shooe and the Hoof, and the latter being well parted, keeping it on with a piece of Leather, and renewing it every three Days; and as you see his Hoof grow again, pare him even to the quick, applying the Plaister, and suffering him to run in soft, tho' not in dirty or miery Ground.

For the Canker in the Head, a Remedy.

When you find by the rawness and yellow matter that this Grievance has seized your Horse; to remedy it before it grow desperate,

Take a Pint of Olive-oil, of *Burgundia*-pitch three Ounces, and an Ounce of washed Turpentine; put them all into a Pipkin, and mix them together over a gentle Fire; and when they are mixed, add an Ounce of Verdigrease, and boil them up to the thickness of a Salve, ever keeping the matter stirring; and making a Plaister apply it to the Canker, according to the advantage of the place where it is situate, having first rubbed off the scurf or scales; and if so it happen to be in the Nostrils, having washed it with a Spunge at the end of a stick, dipped in Salt and Vinegar, to cleanse it, wet the Salve, and dipping a Feather therein, anoint the place grieved with it when warm, and capable of sticking by the like application.

For the Mangey, or dry Scurvey, a Remedy.

Having cleaned the place by scraping off the scurf or scabs, that it may lie open to the operation of the Medicament or Application,

Take a quarter of a Pint of Strong-beer, with two Ounces of the Oil of Turpentine, and well mixing them by shaking in a viol glass; anoint the place grieved with a feather, tying up your Horse to prevent his unruliness, during your so doing, and till the sharpness of its operation be over, with a strong cord to the rack; then blow upon it Powder of Bolearmorick, and bind the Sorrance gently with a cloth: this you may repeat once a Week, as often as you see occasion; and when it heals, which will be signified by the returning of the Hair, then you may supple it with Ointment of Marshmallows, and wash it with water wherein Charvil has been concocted or boiled.

The Vives, *and their Remedy.*

The *Vives* are a troublesome Distemper or Grievance, that frequently happens under the Ears of a Horse, and many times endanger his Life: For which, in the first place, let your Horse Blood in both the Neck-veins, then placing your Bernacles upon his Nose,

to make him less sensible of pain, shape an Iron as you see the form in the Margin, suffering the edges to be about the thickness of a half Crown; heat it red hot and scarifie the place upon the middle of the swelling, from the root of the Ears, downwards to the lowest part thereof, in the form annexed, till you perceive the Skin of a yellowish colour, then desist; and having passed over the place with a cloth, anoint it with Oyl of Roses, or, for want of that, fresh Butter or Hog's Lard, keeping the place supple by often repeating it; as also the Horse in a warm Stable, with good Diet.

For Swanking in the Back, or a Strain in the Kidneys, caused by indiscreet Riding, or Over-burthening.

Your Horse being under these circumstances, mix well together two ounces of Nerve-oyl, and the like quantity of the Oyl of Turpentine, over a gentle fire; and having a Sheep's skin newly stripped off, rub it with a Brush of Cloth all over the fleshy side, and clapping the Skin upon the Horse's Back, especially where the Grief is, bind it on with broad Sursingles very strait, bracing it with a Crupper behind and Straps before; and give the Horse the Juyce of Pelletory, sweetened with Sugar candy, half a Pint warm, in a Pint of Ale.

For Pains or Foulness in the Reins or Kidneys, an excellent Scowring.

Take Treacle jean, two Ounces, and Rhubarb in Powder half an Ounce, with an Ounce of the Juyce of Hysop to qualifie them: put these into a Pint of Beer or Ale, when very hot, and give it to the Horse fasting.

A present Relief for an Attaint, or Over-reach on the Heel, or the like.

This Mishap cometh to pass when the Horse with the Toe of his hinder Shoe strikes the Heel just at the setting on of the Hoof, commonly called the Over-reacch; and if not timely regarded, may prove dangerous, for being not only a breaking of the Flesh, but a strong bruise, it sometimes by its rankling perishes the Sinews, or otherwise renders the Horse lame or disabled: and in this case clip away the Hair, and the batter'd Skin or Flesh, which you will find hang loose and useless; and having so done, wash the dirt out of it with Water and Salt, after which anoint it with Neats-foot Oil, or Mutton-suet; and then dip a wad of Flax in the Whites of Eggs, and bind it hard with a list or soft string to the place, and renew it will you find the Sorrance healed, which will be in a Week or thereabouts.

For the Water-Falcion, a Remedy.

This Distemper is occasioned by the Horse's unwholesome feeding in low wet Ground, where the moisture is great, insomuch that with the Grass the Horse takes in extraordinary quantities, which frequently occasions soft swellings under the Belly and Chaps.

To Cure this, Work a piece of Iron in the fashion of a Fleam, and having heated it red hot, strike it through the Skin of the Swellings, and the contracted Humours will flow thence, being an Oily-water of colour yellowish, and sometimes greyish; and then there needs no more than to wash it with Chamber lye, as hot as can be well endured, having mixed it with the infusion of Tobacco-stalks, and Powder of Bole-armorick.

A Cure for the Sorrance, called the Ring-bone.

This Ring-bone is an Excrescence, generally growing upon the Instep, just above the Hoof, on the fore part of the hinder Leg, and is many times as big as a Pigeon's Egg: To remove this,

Tye up the contrary Leg of your Horse, and strike with a sharp Bodkin, according to the form in the Margin, five or six holes in the Ring-bone, at the edge of it, suffering the holes to be of an equal distance, and put into them Arsnick or white Mercury, beaten into fine Powder, and with the Skin of Mutton-suet bind the Sorrance up for the space of a day and a night, and it will eat away, by its corrosive quality, the foundation, so that the Ring-bone being anointed with Supplements, will fall off or crumble away.

For the Ives a Remedy.

This Distemper is found to grow like a Roll between the Neck and hinder part of the Jaw-bone, and is of dangerous consequence if it ascend to the roots of the Ears.

The speediest remedy for these is to let the Horse bleed well in the Neck-vein, then take Pepper, Hog's Lard and Vinegar, of each half an ounce, with a spoonful of the Juyce of Savin; make them up as thick as may be, put one half of them into one Ear, and the rest into the other, stop them with Lint, stitching up the Ears so fast, that he cannot shake them out for the space of twenty four hours, and they will distil into the Head an operative quintessence that will dissolve the Swelling.

To take off the Film or Skin from a Horse's Eye.

To remove this Obstruction to the sight, Take a piece of lean hung Beef, or other salt Beef; dry it in an Oven, to that degree, that it may be reduced to Powder, and do the like to a stick of Licorish, so take of them an equal quantity, and a third part of burnt Roach-allom: mix them well, and each Morning blow with a Quill about a Penny-weight into the Horse's Eye, drawing the Lids together, if he will suffer it, the better to keep in the Powder; and in so doing, every other day for five or six Days together, the Film will vanish. This likewise will remove the Pin and Web.

A Mallender, the Remedy to Cure it.

Having rubbed him well with a Cloth, mix Soap with red Mercury precipitate; and having anointed the place grieved therewith, take away the Hair; and having four Days successively anointed it in the same manner, afterwards use mollifying Oils or Ointments to take away the Heat of the former Unguent; and then wash the place with Urine or Vinegar till it be restored.

For the Palsie or Apoplexy, a Remedy.

These Distempers are occasioned by the Nerves and Sinews, as also the Brain being afflicted with bad Humours or ascending Vapours; and the Signs are the stiffness of the Neck and hinder parts, the hardness of his Flanks, and the dullness of his Eyes. To remedy which,

Take the Oil of Peter, and chase it into those parts you perceive to be afflicted; force it for its more speedy penetration, with an Iron indifferently hot, and, after that, give him half a Pint of Pennyroyal-water, sweetned with Sugar, and cover him up warm: you may, if you can oblige him to lie down, cover him with the reeking Litter, and gently twist a Thumb-band of the same all over his Neck.

A Farcion in the Head and Neck to Cure.

For this Distemper, proceeding from corrupt Humours, the Neck-veins must be breathed: Then mix the Juyces of Housleek and Hemlock a like quantity, not exceeding two Spoonfuls; adding a Spoonful of Olive-oil, and dividing them into equal potions, put a half into each Ear, stopping it in with Cotton or Lint, and tie up the Ears for twenty four Hours, giving him at the end of three Hours a warm Mash, with a few Coriander-seeds in it.

A Lineament to cleanse a Wound, new or old.

Take Elder-roots dry, and beat them to Powder, and boil the Powder with Honey, and a little Allom-water, and make a Pessary or Lineament; and dipping it therein when blood-warm, wrap it round your Proble, and gently cleanse the Wound; washing it likewise with Water, and it will cause it to fill with flesh.

For Kibed-heels, commonly called the Mules, *a Remedy.*

These Sorrances are no other than dry Scabs breeding upon the Horse's Heels, and so inward to the Fetlock, in long chinks, chops and creases, *&c.* and the occasion is going in wet and dirty ground, and then heated without any regard or good looking to, which makes even a good Horse, when so afflicted, stiff and unfit for service: To remedy which,

Take calcined Tartar, and dissolve it in Water; and when it is congealed in the nature of Salt, mix it with Soap, and the Oyl of Tobacco, and with it anoint the Sorrance, washing it before and after with strong Beef-broth, and in four or five days, with this continuance, they will be well, especially if the Chops or Rifts are not exceeding deep.

The Quitter-bone; what it is, with its Remedy.

The Quitter-bone, is either occasioned through the fretting or corroding of Gravel under the Shooe; or by an unfortunate or careless prick with a Nail left unregarded against the Sinews, which causes the Humour thereby contracted to move upwards, and break out in a round hard Swelling on the Corronet of the Hoof, and in four or five days will it break and send forth Matter from a deep hole like an Ulcer, and is in it self very dangerous and troublesome; the best way to sink or destroy it, is to Cauterize it with a hot Iron, made in the fashion of a Half moon, as thus: Let it be done almost round, and quite cross the middle overthwart; then prick it full of holes an equal distance from each other, and put in Arsenick or Mercury, covering it over with Soap or Butter, binding it down with a Lineament for the space of two days; and then, if you find it black, the effects are wrought by the Poyson, insomuch that you may venture to take away the Core, though it hang a little by the Grisle, and lint the hole with a Lint dipped in Verdigrease and Honey well boiled together, anointing the parts about it with Hog's Lard, as a Supplement till it be well.

An excellent Cure for the Blood-spavin, &c.

Take up, and knit the Vein above the Grievance; and having divided it, take of Linseed a Pint, bruise it well, and fry it in a Pan with new Cow-dung; and add more, four Ounces of Hog's Lard, and two of the Juyce of Hemlock; and so in the form of a Cataplasm apply it, renewing it every day, and it will cause the Sore to come to a head, and then, by breaking it, brings away the Corruption.

For a Neather-attaint, or Over-reach in the Pastern-joint.

This Sorrance is known by a little gellied Bladder on the hollow of the Pastern-joint, not much differing from a Wind-gall; which though many times not to be seen at a distance, yet it may be found and known by feeling. To cure or remove this,

Take a small Cord or Lift, and rowl it somewhat strait from the Knee to the Neather-joint, and then with a Fleam let out the corrupt Matter: which being pressed out,

Take the Whites of four Eggs, a handful of Bay-salt, and two Ounces of the Juyce of Hemlock: mix them well together, and dipping a Ragg into them; or rather a Wadd of Flax, having first unbound the String, lay it on the place where the Swelling was, and bind it softly on; and so continue to renew it for the space of four or five days.

For a putrified Frush, the Cure.

Having well washed and cleansed the Foot with Man's Urine, take of beaten Allom a Pound, and put it onto a Quart of the same Urine; and gathering a good quantity green Nettles, dry them so that they may without difficulty be beaten into Powder: do the like by Pepper; and when you have, after Travel, or any Exercise, washed the place grieved with Urine, in which the Allom is dissolved, blow the Powder upon it, and so bind it up: And by doing this frequently, you will find the effects answerable to your expectation.

To dissolve the Humours, and thereby anticipate Diseases.

Take Sage, Rosemary, Wormwood, the Barak of the Root of an Elm, or the Leaves of the Pine and Worm-wood, of each a handful; stamp or thread them, and then boil them in the Oil of Linseed, till they, being pressed, become the thickness of an Ointmnet; and with it, as hot as may be, chafe and rub the place where you perceive the Humours to settle or begin to draw together: and by often so doing they will begin to disperse. Figs and Salt boiled to Gelly, with the Juyce of Nettles and Elder, have in many cases the same Effects.

To soften any hard Swelling, or contracted Hardness.

Take of Neats-foot-oyl a Quart, of the Juyce of Coleworts half a Pint, of the Marrow of Hog's Feet two Ounces, and an Ounce of the Oil of Cyprus, with half a handful of the Roots of Mallows bruised: boil or heat them over a gentle Fire till they incorporate and become an Ointment; then, the Roots being taken away, put it up into a Gally pot, and as often as you see occasion use it hot to the grieved place.

For the Splint, Wind-gall, or Bladders of Gelly, in or about any of the Joints subject thereto.

Take Bees-wax a Pound, Per-rosin half a Pound, *Galbanum* two Ounces, *Sal-armoniack* an Ounce, *Costus* three Ounces, *Myrrh Secundary* a Pound: bruise and melt them together till they are well incorporated; and so, being made into a Salve or thick Oyntment, use it Plaster-wise, by applying it to the Grievance, and you will soon find it effectual.

To cleanse any putrified, or other Sore, the best way.

Take Salad-oil and tryed Hog's Lard of each a Pound; Turpentine and White-wax of each four Ounces, and six Ounces of Allom-powder, and a quarter of a Pint of the Juyce of Rue: make them into an Ointment over a gentle Fire, and dress the Wound therewith, as you see occasion, and it will not only cleanse it, but fill it with sound flesh.

A Horse Planet-struck, how to cure.

This Distemper takes a Horse's Limbs away on a sudden, so that they remain for a good time in the same posture they did at the time of the seizure, he not being able to move them: And this, though it is by the English called *Planet-struck*, and by the French *Surprius*, yet it is no other than the effects of heat and cold; and whether of these it is, may be thus perceived:

If it be cold, then it is discerned sometime before by his snuffling and rattling in the Head, which denotes that cold phlegmatick Humours do assault the Brain: And if from heat, then it may be perceived by the dryness of Tongue, the scorching of the Breath, clear breathing, and the like, then is the Malady in the Blood, composed of Crudities and gross Humours; for the first, anoint his Temples with the Oil of *Petrolum*, and give him an ounce of *Leserpitum* in a pint of Canary, and half a pint of Olive-oyl, as warm as may be: and for the latter, having blooded your Horse, give him Water and Honey, with an ounce of *Leserpitum*, and two ounces of Melion-seed bruised to Powder, and let his Diet be moderate, especially if his Body abound with gross humours, that by a spare diet they may waste and consume; though sometimes indeed by extream fasting this Distemper happens, and then good feeding, though by degrees, is the best remedy.

For the Poze, or excessive Cold, a Remedy.

Take Conserve or Ellecampane, or else the Herb, bruise or dissolve it in a pint of Mallage; then add an ounce of beaten Ginger and Powder of Rosemary; and having well warmed them over a fire, give them the Horse to drink, and so continue to do every other morning for a Week.

To remedy or cure Hideboundness in a Horse.

This disorder of the Body you shall know by finding the Skin of your Horse cleave to his Ribs and Back-bone, so that you cannot without much difficulty take it up; and this proceeds from a pining or wasting, by reason of some inward Distemper of Body, or by having been bad kept, hard rid, or suffered upon a Heat to stand much in rainy weather, and be afterward unseasonably dryed: To remedy which defect, and render his Skin loose and pliable, so that he may thrive and recover his Strength,

Take of Cummin and Anniseeds, each two ounces, the Powder of Licorish an ounce, Flower of Brimstone half an Ounce, and Oil of Roses a quarter of a pint; mix these together, and heating them well in a quart of Ale, give them the Horse to Drink Morning and Evening for the space of a Week, and the remedy will prove its sufficiency.

For a wet inward Cough, a Remedy.

This Disorder proceeds from the Horse's being too much in damp soggy Airs, whereby gross Humours are ingendred; which converting to Rheums, fall in such abundance upon the Lungs, that they, as it were, stifle and overwhelm them in such a manner, as renders them almost uncappable of performing their Office; so that the sound by that means seems to be inward: Now to remedy this, and prevent the danger,

Boil a Peck of Barley to a Mash or Pulp, then add to it Licorish-powder two Ounces, Anniseed and Carroway-seeds, of each an Ounce, sliced Dates, or blew Figs, half a Pound, Sugar-candy a Pound, Turmerick three Ounces, and two Roots of Garlick, with a Quart of Olive-oil; and when they are boiled to a Mash, press out the liquid part between two Cheese-fats, and give him a Pint of it hot for six Mornings together, and soon after exercise him smartly, the Weather being dry, and the Ground good.

For a dangerous Cough, commonly called the dry Cough.

This Distemper proceeds from bad Feeding, or unseasonable Labour, which causes the cholerick Humours to abound, and fall upon the Lungs, in a hot, tough, yellow flegm, which clings them up as it were, so that the Horse cannot breathe without Pain; and tho' he seldom Coughs, yet when he does, it is performed with a hollow sound, and much violence: Now to remedy this,

Take a handful of Camomil, and the like quantity of the Herb Mellior, two Ounces of Licorish-powder, and three Ounces of the Conserve of Red Roses, a quarter of a Pound of Honey, and two Ounces of Allom; boil them in Water, wherein four Ounces of Camphire has been dissolved, and give him the liquid part to drink it as hot as he may well endure it, keeping him for two Hours after fasting, and beware he catch not Cold.

For the Yard of a Horse falling, a Remedy.

This happens to a Horse when he is grown feeble, either by over-labouring, or bad feeding, which cause a bad resolution in the Muscles and Tendons, so that they refuse their Office of Support: and in this case,

Take a gallon of Water, boil in it two handfuls of Bay-salt, half a pound of Carrot-seed, a good handful of Mug-wort, and the like quantity of Bay-leaves; then strain out the liquid part, and add a quart of old Mallago, and give him this to drink hot, rubbing his Yard with Vinegar, wherein Nettle-seeds and Burdock-seeds have been concocted.

To prevent the Mattering of the Yard.

This happens to Horses of a hot Constitution, especially after covering; and first appears by the swelling of the end of the Yard, and his being thereby uncappable of drawing it into his Sheath, when soon after you will perceive much filthy Matter issue from thence; at what time dissolve half a pound of Allom in a pint of White-wine, and with a Syringe inject it warm into the Yard, whereby the Yard will not only be scowred and cooled, but the Humours be driven back and dispersed.

Of the Diseases incident mostly to Mares, and known by the name of the Pestilent consumption.

This Distemper happens to a Mare, when she is near her Foaling-time, by reason of a Flegmatick Humour that contracts about the Matrix, occasioned by gross Feeding, and is known by her dullness, pining, and desire to be Laid, and the like: to redress which,

Take a pint of *Aqua vitæ*, half an ounce of Tobacco, and a sprig or two of Spurg-lawrel; boil them together, and then straining out the liquid part, give it her fasting, and it will oblige her to cast out the Mass of Flegm, or at least the cause that disturbs her: But by reason she will be somewhat sickish when she has cast, give her half a pint of Salad-oil, and the like quantity of Canary, and keep her in a warm Stable, with Mashes and good dry Meat a day or two.

How a Mare that is subject to cast her Foal ought to be used.

In this case there is more than one Cause to be taken notice of, which subjects a Mare to cast her Foal untimely, sometimes dead, and sometimes alive; as a hard Wintering, unwholsome lodging, over-riding, sudden strains, or unhappy blows on the back, leaping hedges, or the like; together with too much fatness, or subjection to gross humours, many times to the endangering of their lives: Therefore when you perceive her near her time, bring her into a warm Stable, and

Take an ounce of *Diapente*, an ounce of *Sarsaparilla*, three grains of Musk, and a penny-worth of Fennel-seeds; infuse them into a pint of Muscadel, and give them a heating over a gentle fire, and suffer the Mare to drink the liquid part fasting; then dip your hand in the Oil of Myrrh, and thrust it into her Shape, and give her wholesome Diet and good Litter; or this may be done, with success, at the time of her Foaling, especially if you perceive difficulty therein, and it will be much available in rendring her an easie delivery.

How to oblige a Mare to Cast her Foal.

Now, on the contrary, if you imagine the Foal your Mare goes with, is not worth your rearing, or that she has taken Horse contrary to your desire; then boil a good quantity of Savin in two quarts of Milk, adding an ounce of Rhubarb and a small quantity of Wood-ashes, and when they are sufficiently boiled, strain out the liquid part, and give it the Mare to drink very hot, and then give her a considerable heat; and in so doing two or three mornings, the business will be effected. But consider withal, that you look well to your Mare, least you lose both, for she must for a week afterward be kept in the Stable, and that very warm, and with Mashes of sweet Malt-bran and Barley every other morning.

For the Feavers in a Horse, and how to Cure them, &c.

There are divers Feavers that frequently possess the Body of a Horse, and that at different times, as the *Quotidian, Tertian* and *Quartan,* and these are occasioned by gross Humours contracted in the Blood, that inflame and disorder the Frame, and happen according to the circulation of the Blood, or denomination of Humours.

As for the first of these, it is ever the most violent, but never lasteth long, and most frequently it cometh in the Spring, when the Blood begins to increase, especially to Colts and young Horses.

The signs that fore-run this, are the watering of the Eyes, and a redness as if they were Blood-shotten, short pantings, hot breath, a loathing or leaving of Provender, stiffness in the Joynts, and unwillingness to labour; and if it so happen that it befall him at eight of the clock the one day, you may expect it at four the next day, and the reason that is given, is the ebbing and flowing of the Blood, and its circulation. To rid him of this troublesome Companion, give him, as soon as you perceive it to begin, a warm Mash, and keep him in motion, though in the Stable, for the space of an hour or more; then rub him exceeding well, and

Take two quarts of Ale, a good handful of Worm wood, an ounce of Long pepper, Venice-treacle two ounces, and of Grains an ounce; add to these an ounce of the Flower of Brimstone, and as much Rue dryed and rubbed into Powder; heat it hot, and give it him at twice, about the space of two Hours difference between each other.

The *Tertian* Feaver is much the same in quantity and condition with what I have named, and the Symptoms the same, though it somewhat more than the other participates of the Ague, for it at first takes him with a kind of shaking; wherefore when you observe its approach,

Take of Stone-crop, a Herb so called, two handfuls; bruise it, and strain the Juyce into two quarts of Ale; drop into it then an ounce of the oyl of Myrrh, and an ounce of Ginger beaten into Powder; make them hot, and give them the Horse to drink, sweetned with Sugar-candy, and then rack him a round pace in wholesome Air; but do not sweat him, that he thereby may be endangered, by contracting a Cold, observing to let him drink no cold Water till such time as you find the Fit intirely gone, and that he has settled his Body by eating two quarts of the best dried Oats.

As for the *Quartan* Feaver, it is much of the nature of the two former, only it alters the day, and often continues longer; for if a speedy remedy be not had, it frequently continues, at divers times, for the space of half a Year, or longer; and if this happen in the Fall of the Leaf, it will be necessary to let Blood; which done, give your Horse what is hereafter directed,

Take Oil of Bays an ounce, Colts-foot the Herb, a good handful, Knot-grass-roots, or dryed Lavender, a handful; boil them in Ale or White-wine, and give him the liquid part, ordering him as for the former.

For any Disease and Stoppage in the Liver.

The Obstructions in the Liver frequently happen through excess of Humours, that are not capable of being digested into good and wholesome Blood, clog and hinder the cavities of the Passages, and by that means cause Pains and Sickness. Now, to remove these Humours, known by the dullness of the Countenance, hanging of the Head, often straining, and inward groaning,

Take Agrimony, Camomil, Fumitory, Pursley, Wormwood, Succory, Endive, the Seeds of Lupins, and Flowers of Mayweed, a handful; Licorish, Gentian and Spikenard, of each an Ounce: bruise them well, and boil them in a quantity sufficient of Sider or Perry, and give it the Horse very wam, and let him walk thereupon for the space of an Hour after; and for a Fortnight after be sparing in his Diet, that the Humours by this means may disperse and consume.

For a Botch or Sorrance in the Groin of a Horse, a good Cure.

When by the tumourousness of the Flesh you perceive a Swelling to arise in the Groin of a Horse,

Take Shooemaker's Wax, the white sort, add to an Ounce of it half an Ounce of the Powder of Bithwort-roots, and as much *Amoniacum*, and, over a gentle Fire, make them into a Plaister: which being spread upon a Leather, apply it to the place till the Swelling is ripe for breaking; then lance it, and take out the Putrifaction; after that wash the Sorrance with Water, wherein Allom and Honey have been dissolved, till you find it begin to heal; then anoint it with *Unguentum Aegyptiacum*, and bind it up.

For a general Manginess, the Remedy.

This happens through the Corruption of the Blood, and grossness of Humours, occasioned by over-labouring, over-heating, and bad feeding, or any of these; and may be sometimes catched by Contagion from other Horses; the sign is an extraordinary itching, which you may observe by his scrubbing and the rising of little Knots within the Skin: which being perceived,

Take Verdigrease two Ounces, common Soap a Pound, Oil of Spike two Ounces, Linseed oil a Pint, Red-wine half a Pint: Incorporate them over a gentle Fire, reducing them to the thickness of an Ointment: having first let the Horse blood, anoint him with the Ointment, after the Scurf and Scabs are likewise rubbed off: and so continue to do for a Week, or so long till you perceive the Distemper to cease by the dying of the Scabs, and the coming of good Flesh.

The Barbs, what they are, and how removed.

This troublesome Sorrance happens under the Tongue of the Horse, being composed of two long Bags of Flesh, like Paps of Nipples, growing as they abound with Humour more or less, and hinder the Horse in his feeding, putting him to no small trouble. Now, to remedy it, clip them off close to the Jaw, and

Take Allom an Ounce, Honey the like quantity, Bay-salt a handful, and the Juyce of Mint a quarter of a Pint: dissolve and boil these in a quart of fair Water, and wash the Roots of the Barbs till they heal. Some there are that advise Burning them off; but in my opinion, by reason of their situation, that is neither so easie to be done, nor safe, lest the Tongue-string or small Veins be thereby rendred useless, and consequently the Horse defective in his feeling.

For Blood-shot Eyes, an excellent Remedy.

The Eyes by straining, blow, or super-abundance of corrupt Blood, becoming red and rheumy; so that unless speedily cured, they may turn to further prejudice, if not to blindness. To cure this,

Take the Juyce of a Lemon, the Crumbs of White-bread, Bole-armonick and a rotten Apple: bruise them together, and make of them a Cataplasm or Plaister: Then take the Powder of the Roots of

Mallows, with that of a Crust of brown Bread, and blow into the Eyes, binding over them the Plaister, or rather Poultis, and in so doing three or four times the Blood and Rheum will be driven back and dispersed; but if it be so great that this cannot conquer it, then bleed your Horse in the Temple-veins.

For any Film, Bite, or Blow in the Eye, a Remedy.

Take Copras, that which is white, a quarter of an Ounce, and the like quantity of Verdigrease: beat them to Powder, and dry them well upon a Plate or *Spatula*; and after that take of it to the quantity of half a dram, and blow it into the Eye with a Quill; then close the Horse's Eye a quarter of an Hour, and after that wash it with Eyebright Water; and so continue to do it till all your Powder is wasted, and then you will perceive a brightness in your Horse's Eye, all Grievances being vanished.

To kill Lice, or remove Flies from offending your Horse,

Take of the Flower of Brimstone an Ounce, Quick-silver well killed, the like quantity; the Oil of Spike two Ounces: mix when with the Whites of two Eggs, and then boil them in two Quarts of strong Urine, and anoint the Horse therewith, and it will prevent either the Annoyance, or cure it when contracted.

To rid a Horse from any foulness or disorder in the Body.

Take of Groundsel half a handful, red Sage the like quantity, Smallage and Wormwood each a handful: shread them small, and boil them well in a Pint and a half of Ale, into which put a quarter of a Pound of fresh Butter, and an Ounce of the Powder of Mechocan, give the liquid part to your Horse to drink, as warm as possible, and feed him with Mashes for a day afterward.

For an extraordinary Bruise or Bite.

Take of *Caliminaris,* quenched in White-wine, two Drams; an Ounce of the Juyce of Housleek and two Ounces of the Seeds of Mallows, with an Ounce of *Venice*-Treacle, make them up into Balls as big was Walnuts, and give them the Horse in a quarter of a Pint of Salad-oil; and at the same time apply a Plaister of Hemlock and Barrows grease well stamped and mixed together, and this being done for a Week together, will work strange effects in relation to a cure.

For a Horse that is troubled with the Strangling, a Cure.

Take of Elder-buds, or the Bark of the Root of that Tree, a good handful; Wormwood, and the Herb Mercury, or each half a handful, and as much wild Tansey: boil them well in two Quarts of Vinegar, and give them the Horse, that is, the liquid part as hot as may be fasting.

To Remedy the Swelling of a Horse upon having eaten any infectious thing in his Grass or Provender, that may, if not remedied, prove dangerous.

This Grievance is known by the slavering of the Beast, the staring of the Eyes, and the rising of the Belly, the beating of his Flanks, and a cold Sweat: which perceived,

Take of the Juyce of Rue one Pint, two Quarts of Milk, and a Pint of Olive-oil: Boil them together till a third part be consumed; and then sweetning it with brown Sugar, give it the Horse.

For a Heart-burning or Wasting occasioned thereby.

Take the Juyce of Wood-sorrel or Field-sorrel a Pint, Allom-powder two Ounces, Harts-horn an Ounce, the Seeds of Pomegranates two Ounces, and Spring-water a Quart; boil them well together, and strain out the liquid part, give it the Horse as hot as may be.

To prevent Staling Blood, a Remedy.

Take of Ale a Quart, the Roes of two red Herrings, and three or four Cloves of Garlick: Boil them together, and give them the Horse Morning and Evening, that is, the liquid part.

Another excellent Remedy for the Farcy, vulgarly called the Fashon.

Take Rue, Garlick and Cloves, of each half a handful: bruise them well, and boil them in half a Pint of *Aqua-vitae*; then dip Wool or Lint into the liquid part, and stop it into the Horse's Ears, binding it in: after that bruise the Bark of the Elder, and making an Incision in the Forehead, raise the Skin with your Pegging-horn, and stop it under the Skin, being first dipped in Oil or Rosemary; then give the Horse a Dose of the Juyce of Liverwort, mixed with half a Pint of Canary, and after that warm Mashes.

An approved Cure for any Sinew-strain or Over-reach.

Take Oil of Bays an Ounce, Linseed oil two Ounces; put them into half a Pint of *Aqua-vitæ*, and being well incorporated, add Wine-vinegar half a pint, and boil them to the consumption of a third part; then with your hand chafe it in, or with a warm cloth swathing it afterward to the best advantage, and in so doing you will soon find the effects.

Diseases in the Hoof how to Remedy; and first for a Horse, that upon sundry occasions is apt to cast his Hoof.

The cause of the Hoof's falling off is various, for sometimes it proceeds from Gravel, another time from the prockling of a Nail; and in these cases if cleaves downward as the Humour settles; but if it happen by any Grievance on the top of the Hoof, as the Quitter bone, and the like, then must you look to the top of the Hoof; and when you there perceive it begin to divide from the flesh, or to open at any part, then take off the Shooe, open the Hoof, and pare it ats near as may be to the Sole; after which steep the Foot in Neats-foot-oil, and the Juyce of Hemlock, in which Allom has been dissolved, then make an Oyntment after this manner:

Take of Virgins-wax two ounces, of Verdegrease an ounce, of Per-resin three ounces, and Hog's Lard a quarter of a pound; to these add the Juyce of green Tobacco a quarter of a pint; make them into an Ointment, and pour it into the divided place; which done, bind it up close with a thick Linnen-cloth; and in so often doing, the flesh will be obliged to cleave to the Hoof, and render it firm, unless it be too far gone before you attempt it.

Hoof-bound; what it is, and how to remedy it.

This Grievance proceeds from some defect in the Hoof, or harm taken in the Colt-age, whilst it was tender, or by the falling down of a Humour, and the Symptoms are, the flesh, growing over it more than usual, the straitness or narrowness in disproportion to the Leg, and the sound of hollowness being struck with a Hammer or other material Instrument, which being well understood,

Take Hog's Lard a pound, Soap the like quantity, the Juyce of Baulm a pound, Bay leaves a handful, the Juyce of Rue a quarter of a pint; incorporate or concoct them over a gentle fire, and steep the Hoof in the liquid part, for the space of an hour every morning; then dip a cloth in the Oil of *Petrolum*, and bind it about it.

To soften or harden the Hoof, the best way.

This Experiment is fitting to be known by all Farriers, not only for the advantage of Shooing, but for the Travel, more or less, of the Horse. Wherefore in the first place, if you perceive the Hoof to be hard and britle, standing out, uncomely, then try it with a Butress, and if you find it brittle, and not plyable to be pared or cut; then

Take of Lime unflaked an ounce, common Soap the like quanity; dissolve these into a Lye made of Ashwood-ashes, and having prepared a Cataplasm or Poultis of Groundsel, Marshmallows, Smallage, Succory and Sallendine fryed in the Lye for the space of an hour, as warm as may well be endured; clap the Poultis to them, and stop the bottom of his Foot well with Flax dipped in Tar.

To harden a Hoof as occasion requires.

If by going in moist Ground, or moist Feeding, the Hoof happen to be softened so that it will not well bear a Shooe, or be prejudicial in Travelling; then to render it capable for either,

Take the Coles of burnt Leather a pound, the Water wherein Lime has been flaked, and hot Flint-stones quenched, two quarts; add to them a quart of Lime-juyce, and a pound of Bay-salt, and in the liquid part either let the Horse stand, or with it bathe his Hoofs; after which, Take Oil of Tartar, or that of Brimstone, and anoint them, binding a cloth over them, and suffer the Horse to stand dry: and by thus often doing, you will by experience, find your expectation to be answered.

For any Hurt or cankerous Sore in or on the Hoof.

Take, if the season permit, a pound of Black-snails; if not, other Snails of any kind may serve the purpose; of Burdock-roots sliced the like quantity; Oil of Camomil four ounces, and Olive-oil a quart; boil them together till they are pliable to be laid Plaister-wise to the place grieved; to which, after you have washed the Sorrance with water wherein there has been Elder boiled, applying them, and supplying those first laid with fresh every day, till you find the effects.

To oblige a Horse to carry his Ears well, the way.

This defect happens either for want of a true proportion, or by a defect of the Nerves which refuse to administer so great a supply of strength as may support the Ears as they ought to stand for the sake of Ornament. Now to strengthen the Nerves or Sinews to that degree,

Take of Bear's fat an ounce, Comsory-juyce the like quantity, Oil of Amber a dram, the Root of Black-hellebore, beaten into powder, an ounce, and with Bees-wax half an ounce, and the Oil of Roses half a pint, make these into an Ointment, and anoint the roots of the Horse's Ears as hot as may be well endured, repeating it for a week together.

For the Grievance called the Frounce, *a Cure.*

This Sorrance is also called by some a *Cameroy*, and is no other than small Knobs or Bladders on the roof of the Mouth, or upon the tongue; and the cause, for the most part, is the Horse's feeding in wet or low Marshes in frosty weather, or by eating unsavory Hay, in which Vermin have pissed or dunged; and further, some hold it to proceed from the Horse's licking up unsavory things tending to a venomous quality. To remedy which,

Take Vinegar and Bay-salt, with the Juyce of Save, and with them, after having let the Horse Blood in the Veins under the Tongue, rub the place grieved till the Knobs or Pimples bleed; and by often dong so they will disappear.

For a Heat, which sometimes occasions a breaking out in the Mouth and Lips, the Cure.

This Disorder is accompanied with dryness, and proceeds from the heat of the stomach, by surfeits, over-heating, or a consuming quality; and if not timely taken notice of, produces the infectious humour that creates the Canker: to redress which, Bleed the Veins in the Lips, which you may cause to appear by bending them the contrary way; and then wash them with Salt and Vinegar, giving the Horse water to drink wherein Colt's foot has been boiled, or Fenugreek-seed, with his Provender.

Wolf-teeth, what they are, and how to ease their Pains the ready way.

These Teeth are too commonly growing in the upper Jaw next the Grinders, which many times occasion such pain that the Horse is frustrated in his eating, by being obliged to let his Provender fall out of his Mouth; and the cause they are mostly subject to is this, by reason they have frequently a hollowness within subject to receive the Rheums that settle in the Jaw; to remedy which, either draw them, or launcing the Gums that they may bleed; wash the Mouth each Morning with Hysop-water and Allom, the latter being dissolved in the former.

To stanch any Bleeding, a speedy way.

If by occasion of Wound or Sorrance your Horse happen to bleed excessively, so that thereby, if not timely stopped, he may be weakened, or other ways endangered;

Take the Wool of a Hare or Coney, dip it in Vinegar, and then strew upon it the Powder of Calcined Egg-shells, and apply it to the place; or you may for want of the former, dip it in Nettle-juyce or Bay-salt, or apply it to the Wound or Sorrance a Poultis of Hemlock and the Bark of Elder-root.

To supply the defect of the falling of the Crest, a Remedy.

This uncomeliness in a Horse is the leaning of the upper part of the Neck, on which the Mane grows, to one or the other side, which is caused by a weakness of the Sinews or Nerves, through a contraction of Cold, or some flegmatick Humour there engendred; or, on the other Hand, upon the rising of the Flesh through extraordinary Fatness. To remedy which,

Take the Oil of *Petrolum* two Ounces, Linseed-oil half a Pint; and having well washed the Crest with Water wherein the Roots of Mallows have been boiled, and Allom dissolved, anoint it with the Oil; and in so continuing to do for a Month, the Skin will contract; and more especially, if upon every Application you clap two broad pieces of Deal, or other Board, on each side; and so bind it up in due order, and let the Horse blood in the contrary Neck vein.

To cure Manginess, or the like Disorder, in the Crest.

Take Hog's Lard a Pound, Verdegrease four Ounces, Flower of Brimstone four Ounces; add to these a Pint of Beef-broth very salty, and dissolve what is to be dissolved therein: Then having rubbed off the Scabs and Scurf till they bleed, wash the place grieved therewith as hot as may be well endured, for a week together; and lay after that a cloth dipped in green Ointment thereon. This will also hinder the Hair from falling off, especially the former.

The Navel-gall; what it is, and its Remedy.

This Grievance is no other than a bruise or hurt with an unfit or uneasie Saddle, on that part of the Back that is opposite to the Navel, and for that cause only is so called; it is known by a soft swelling in the place bruised, and its Cure is as followeth:

Take the Whites of two Eggs, an ounce of Copras, two ounces of the Oil of Bays, and of Marshmallows, Smallage, Groundsel and Camomil, each a handful; stamp them in a Mortar, and pour the Liquids to them, by which means make them into a Poultis, and frying them, lay them as hot as may be to the place grieved.

For a Sitfast or horney Excressence under the Saddle, whereby the Horse is disabled from carrying it in good order as he ought.

This Sorrance appears like a piece of old Leather upon the Horse's Back, and is commonly the relick of some old Bruise of desperate Saddle-gall not well cured, and obligeth the Skin to stick fast to the flesh. To cure which,

Take Oil of Vitriol, and anoint the place till it has loosned the excressence: after which take it off by incision, and anoint the place with Verdigrease and Hog's Lard hot, and wash it after with the Juyce of Mint, till it be new skinned; and to make the Hair come, wash it with the Juyce of *Cardus Benedictus*, or that of Hemlock.

For any Knob or Wen near the Saddle-skirt, or the Sides of the Horse, a Remedy.

To remove this, Wash it first with hot Wine-lees, and afterward bathe it with Oil of Camomil, or Marshmallows; then to draw it to a head, lay on a Plaister of Stone-pitch and Turpentine; which being done, launce it with your sharp-pointed Fleam, and apply a Plaister of *Oxecrotium* to draw out the Putrefaction, and then with Hog's Lard supple it; and if the Sore be deep, Tent it with a Linnen Tent, dipped in Bees-wax and Honey melted together.

For Weakness in the Back, a strengthening Remedy.

This happens frequently through Coldness or Watry Humours afflicting the Sinews, or in gelly'd Water settling in the Joynts, or by his too often covering the Mares, or such like. Now to corroborate a Horse defective herein,

Take Horse-radish-roots a pound, Bay-leaves two handfuls, the Bark of Elder the like quantity; boil them in Man's Urine, and with the liquid part bathe the Back as hot as is convenient, giving him for diet Oats and Splent-beans, and each morning, fasting, a Ball made in this manner:

Take Licorish-powder two Ounces, Hart's Horn beaten to Powder an ounce, Fenegreek and Hysop seed stamped, of each two ounces; wet them with as much Mallago-wine as will make them up into Balls about the bigness of Pigeons Eggs.

For the Swelling in the Cods of a Horse naturally, or occasioned by any Bruise, &c.

Take Dill seeds or Fennel-seeds an ounce, the Juyce of Orpin a quarter of a pint, Bole armorick two ounces, the Juyce of Garlick the like quantity; make these, with Hog's Lard, into an Ointment, ad anoint the place grieved as warm as may be well endured.

For Burstenness, or Rupture in a Horse.

This Grievance is the breaking of the Rim or Film that holds up the Bowels from falling into the Cods, and either happens by overstraining in riding upon a full Belly, or the Horse's leaping beyond his strength, and is by most Farriers held incurable. But since it so happens that a Bursten Horse may, notwithstanding, in some measure be fit for Service, I shall give such Directions as may the better enable him for it:

Take your Horse to whom this misfortune has befallen, and fasten Ropes, with running Nooses, to his Feet; then putting Ropes through them, cast him upon soft Straw, and draw him up by casting the Ropes over a Beam in the Barn or Stable; then with your hands gently separate the Bowels from the Stones, putting the former in their proper place, and retaining the latter; withal, anointing the Cods with Hog's Lard and the Juyce of Endive, gentle chafing them till they begin to shrink up; then with a soft List tye the Cod as near to the Belly as may be, and cut out the Stone where the Film is broken, sewing up the Cod with thread dipped in the Oil of Turpentine; and so keep your Horse for the space of a fortnight in a warm Stable, with heartning Die, anointing his Cods with Hog's Lard and Allom-powder melted together over a gentle Fire.

The Art of discovering hidden Griefs, or Ailments in a Horse, and from what they proceed.

The better to be the least couzened in Horse flesh, is to see one lead him before you in a string to several paces; and though you cannot so know his Grief or Ailment, yet if you will observe how he takes up his Legs, the defect will appear by his favouring one more than the other; but if nothing of that appear, then back him and ride him at all paces, so that you or your Friend, who stands by, may be convinced by his going, how he stands affected. Now if so it happen he sets out stiff, and handles his Legs confusedly, then does his Grief proceed from cold settlements of watry Humours in or near the Joynt; for which, chafe his Legs morning and evening with *Aqua Vitæ* and Oil of Turpentine; but if it proceed from Heat, as Surfeits and the like, then he will halt the more he is ridden or chafed, by being heated, and then anoint his Limbs with Neats-foot-oil, the Marrow of a Swine, or wash them with Water, wherein Copras has been dissolved, and look he be kept in a clean warm Stable.

The Bone-spavin, what it is, &c.

This Excressence or Sorrance is a Knob of bony substance growing under the Joynt, on the side of the Hoof, as big sometimes as a Pullet's Egg, and either proceeds from the too early pressing the little Bones in those parts, by hard Labour, or from the settlement of bad Humours, fed by the Master-vein that descends to those parts. This many times is so troublesome to a Horse that it makes him go down lame. Now the best way to remove and cure this Sorrance is this:

Take a Pen knife and lay open the Skin about the bony Excressence; and having a Chisel or Knife for that purpose, strike off, or pare away the Bone till you come as near as may be to the quick; then having ready a Plaister of Bees-wax and Verdigrease well mixed by melting, bind it on, but so that it may not afflict the Vein, and renew it every other day, for the space of a Week; then with Hog's Lard cleanse and anoint it for two or three days more, and finally, wash it with Vinegar, and bind it up that it may heal. For want of Verdigrease, you may take the Powder of burnt Allom.

For a Haw in the Eye.

This happens most commonly by a blow, or over-riding, and greatly impairs the sight: To remove which,

Take burnt Allom, and the Powder of Juniper-berries, and blow them into the Eye, keeping it close for a quarter of an hour; and so by often doing, you will perceive the Haw loosned by its wrinkling up, then with your Nails take hold of it, and draw it forth, or if you cannot do so, continue the Powder till it is eaten off.

For the falling out of the Fundament, a Remedy.

The falling out of the Fundament often happens by an extraordinary Laxativeness, or a Coldness contracted in the Bowels, which creates a slimy Matter; and in this case anoint it with Oil of Spike, and sprinkling on it the Powder of Turmerick, put it up in order, and binding a string about the Tail, bring the string between the Legs, and by fastning it to a strap cross the Shoulders, keep in the Fundament, suffering the Tail to continue so fastned for the space of twelve Hours; and this method may be sued in case of the Womb of a Mare's falling out, *&c.*

To preserve a Hoof from decaying.

Take half a pound of Tar, a quart of Vinegar, half a pound of Hog's Grease, a quarter of a pint of the Juyce of Garlick, two ounces of *Olibanum*, and as much Bees-wax; boil them together till the moisture be so far consumed, that it becomes the thickness of an ointment, and with it at seasonable times anoint the Hoof, and dipping Flax into it, stop the hollow part, if you find any defect there, or suspect any will happen.

Interfering, what it is, and the Remedy.

This happens when by the unevenness of a Horse's steps, or the bad shooing, he cuts his fore Fetlock joynt on the inside with his hinder shooe; which by its not being well looked to, many times festers and becomes a sore Scab: To cure which,

Take fresh Butter half a Pound, Rosin the like quantity, Nerve-oil half a Pint; melt them together, and when they are of a thickness, spread a Plaister, and lay it to the place grieved, supply it before with Hogs-grease, or the Oil of Camomil.

A false Quarter, what it is.

This is a defect in the Hoof, in such a manner, as if a piece was put in, having on either side it a Seam or Rift, which frequently obliges the Horse to halt on that part, and is generally caused by some prick or hurt when a Colt. To remedy which, at least to give the Horse ease,

Take off the Shooe, and pare the Hoof, on that side as much as may be, and then supply the defect with Toe, dipped in melted Turpentine and Bees-wax, not over-traveling, or using your Horse to dirty ways.

The Melt on the Heel.

This is no other than a dry Scab growing on the Heel, occasioned sometimes by the Horse's standing wet and dry over long or unseasonably; and at other times, through corrupt Blood settling there: To remedy which,

Take of black Soap a quarter of a Pound, Honey the like quantity; dissolve them in a Pint of Vinegar, then add the Powder of burnt Allom two Ounces, and Rye-meal a like quantity; wash the Sorrance well with Water and Salt, and then spread the before-mentioned Materials, and apply them Plaister-wise, having first taken off the Scurf or Scab as clean as may be; and so for a Week together continue the Supplement.

An excellent Remedy for any Strain or Swelling.

Take of *Aqua-vitæ* a Pint, melt into it a quarter of a Pound of fresh Butter, and mix with them a quarter of a Pint of the Juyce of Hellebore: then, with the Liquids, make a Plaster of Rye-meal and apply it as hot as may well be endured to the place grieved: This will likewise asswage any Swelling, or the like Disorder.

How to remedy the Harm done a Horse by unadvisedly and unskillfully letting Blood.

Many times a Horse being let Blood by any unskillful Hand, or suffered thereupon to take Cold, or the Wind to possess the empty Veins which causes Swellings in the Neck, or other Disorders. To remedy it,

Take Sheeps suet half a Pound, the Juyce of Hemlock half a Pint, and four Ounces of the Oil of Camomil: which being made into an Ointment, rub and chafe his Neck with them as hot as may be, Morning and Evening, giving him warm Water to drink wherein Fennel-seeds are scattered, and cover his Neck over with a warm Cloth, giving him gentle heats.

For the Leprosie in Horses, a Remedy, with the Cause.

Take Resalgar, otherwise called Arsnick, and Hogs-lard well tried: incorporate them to an Ointment over a gentle fire; and having drawn the Horse's Head up strait to the Rack, to prevent his Disorders, anoint the place with a Feather, and suffer it for the space of two hours to soak in; and after that boil the Roots of Burdocks in Chamber-lye, and wash it with the Ointment clean away: which done, give the Horse Meat of the best to hearten and encourage him to endurance; and so proceed to do every other day for six days successively.

This Grievance, or dangerous Malady, befalls a Horse by extraordinary riding, and suffering the Horse to cool, and consequently surfeit: or from the rankness of Blood, which produces evil humours, and they not timely let out force their way in Botches and dry Sorrances, which upon dressing must be rubbed off, to prepare the way for the Ointment.

For any Disease in the Lungs, an excellent Remedy.

The Diseases in the Lungs proceed frequently from extraordinary cold and flegmatick Humours, or, on the contrary, from hot Inflammations, caused by Surfeits, or the like; either of which, if not timely remedied, tend to consumption and rottenness, and are known by the working of the Ribs, and beating of the Flanks; but more especially by his coughing weakly, and the slow beating of what has been before mentioned, with other the like signs.

To cure these, Take of the Herb of Mellilot, commonly called the Horse-lung wort; bruise it in a Mortar, and squeeze out the Juyce to the quantity of two Ounces; of Fenugreek seeds and Madder, each an Ounce, with as much of Rosemary-seed, and give him them, the latter being well bruised in a Quart of warm Ale, every other day, for the space of fourteen days, fasting; and, after the Dose, let him have Oats washed in warm Beer, and warm Mashes, keeping him in a close Stable, without exercising him, unless in a fair clear day: Or for want of these,

Take a young Snake, open it, and put into the Belly Rue and Snake-weed; shread, with the Fat of a Hedge-hog, as much as the Belly of the Snake, being embowelled, will hold; and let it be roasted before a gentle Fire, saving the Oil or Dripping that falls from it carefully in an Earthen-pan; and having shaved off the Hair on the Breast, anoint it with this Ointment, chafing it in with your hot hand; and so do by renewing the Ointment as often as you see it convenient.

For the Swelling of the Horse's Legs, the Cure.

The cause of the Swelling in the Legs comes through cold Humours settling therein, or over much riding in foul or dirty Ways, over-heats or over-strains, or by Molten Grease falling down into the Legs; and in this case, having let Blood in the most convenient Veins, as near as may be to the Swelling, to take away the corrupt Blood, then,

Take the Lees of White-wine or Rhenish-wine half a pint, Camomil half a handful, Cummin-seeds an ounce, Wheat-flower two handfuls; boil them altogether, adding in the boiling half a Pint of Verjuice, and so lay them hot as a Poultis to the place grieved, renewing it till you find it draws the Swelling to a head: which being done,

Take Shooemaker's Wax an Ounce, the like quantity of Virgin's Wax; half an Ounce of Bole-armoniack, and half a Pint of Olive-oil; the Yolks of two Eggs and half a Quartern of Honey: beat these well together over a gentle Fire, till you perceive them well incorporated into the thickness of a Salve; and then, spreading part of it upon Sheep-leather, apply it Plaisterwise till the Corruption, by often renewing it, be drawn away: then wash the place with Balm-water, and heal it up with Hogs-grease and Honey, incorporated over a gentle Fire.

The flying Worm, what it is, and how to Cure it.

This is known generally by the name of a Tetter or Ring-worm, occasioned by an extraordinary heat in the Blood, and other foul Disroders, creating a virulent or sharp fiery Humour, and for the most part seizes upon the Rump or Crupper of the Horse, and frequently by not being regarded, turns to a Canker; though indeed it will seize, and so is found to do upon any part of the Body that is subject to fleshyness or abundance of Blood, and cause the Horse to rub himself in an extraordinary manner: and is known by the falling away of the hair, and the Horse's continual rubbing:

To cure this, Let the Horse blood as soon as may be in that part where it has seized him; and then

Take an Ounce of Verdigrease, two Ounces of Burdock juyce, two Ounces of Soot, a quarter of an Ounce of the Oyl of Tartar: and with the weight of all these in Hog's Lard, make them up into an Oyntment, bathing it with it as hot as may be endured: and so continue to do till the Malady ceases to spread, and consequenetly dies.

Excellent Directions for the preventing of Diseases in Horses, at sundry times, &c.

Observe in this case to bleed your Horse the beginning of *April* in the Neck-veins, when the Sign of Life is not on that part; and so, every day in the Month of *April,* give him what I order: As thus,

Take old Rye, not musty, nor any ways disordered, impaired by shrivling up, nor any way foul with Dirt, Lomestones, or the like, take to the quantity of a Bushel; and having sprinkled it with the Juyce of Baum, and again dried it by spreading in the Sun; put it into an Ion-boiling-pot without Water, and there, by perpetual stirring, parch it to that degree that it becomes black and hard: then take it out, and put it into a close dry place, and each day give your Horse a Quart of it, beaten to Powder, amongst his Oats: and so do in the Month of *October,* likewise remembring to let him blood: And by this means his Blood will be so well tempered, that unless some extraordinary matter happen, as exceeding heats occasioning surfeits, too rank feeding, or damp lying, the Horse will undoubtedly be kept in a good temper of Body during the whole Year And the better to Confirm him, give him this Drink as soon as he is let blood, *viz.*

Take of White wine a Pint, infuse into it Cinamon, Cloves and Saffron, of each three Drams, Cassa and Myrrh, of each the like quantity; let them simper over a gentle Fire for the space of an Hour, and then being sweetened with Sugar-candy or fine Sugar, give it him Blood-warm fasting, and keep him warm with a Cloth for the space of two Hours after without Meat.

A Cure for a sore or defective Mouth.

The Sores or Disorders in the Mouth are occasioned either by bad Blood or excessive Colds, creating Rheums and noisome Vapours that afflict the Palate, for there they generally begin; and from thence descending to the Jaws, do in a great measure obstruct the feeding, and hinders the shutting of the Mouth: Now when it happens in the Pallate only, the best expedient is to let Blood in the Mouth, by cutting the third Barr, or as your direction leads you.

The Horse being let Blood, Take of strong old Cheese four Ounces, and a Root of Garlick; bruise them well, and boil them in Water wherein Plantane has been concocted, and with the liquid part wash the Mouth and Tongue of the Horse, as hot as it may be well endured; so continuing often to do, till you find the Grievances to decrease; and if it be so far gone, that this proves not effectual,

Take a Pint of Verjuyce, a Handful of Bay-salt, a quarter of a Pint of the Juyce of Housleek, and a Pennyworth of *Diascordium*; boil them well, and having washed his Mouth with Savin-water, or Water wherein Savin has been concocted, give him the before-mentioned Potions to Drink luke-warm.

For the Mellet in the Heels, a Cure.

Take three Ounces of Castle soap, a Pound of English Honey, Allom two Ounces, and of Lime-juyce or Verjuyce a quarter of a Pint, with half a Handful of Bean-flower; incorporate them over a gentle Fire, and having reduced them to a convenient thickness, bind a part of it with Leather or thick Linnen upon the place grieved, suffering it, without renewal, to continue there for the space of five Days; and between each renewal wash the place well with Beef-broth, keeping his Leg moist and roped, for some Days after.

The Stavers, their Signs, Cause and Cure.

This Distemper is known by a dizziness in the Head, a dullness in the Eyes, and disorderly hanging of the Jaws, and proceeds in chief from corrupt Blood and infectious Vapours that infect the Brain, and consequently put the whole frame out of order. And this Disease but few Horses altogether escape.

To cure this, to let him blood in the Temple-veins or Neck-veins; and having a potion made after this manner, give it him hot, *viz.*

Take a handful of Savin, the like quantity of Rue, an ounce of Rhubarb, and an ounce of Mithridate; give him these Ingredients, the liquid part well boiled in a quart of Spring-water, sweetened with brown Sugar or Molossus.

For the Stone, a very good Remedy.

This Distemper is occasioned by gross Humours, which settling in the Reins and Bladder; do in process of time, by the help of Heat and Moisture, grow to a hardness, and so obstruct the passage of the Urine; and by grating those tender parts cause exceeding pain and disorder to the Creature so afflicts. To dissolve or remove the Stone so contracted,

Take the Roots of Nettles, Parsley, Fennel and Sperage, of each four ounces, of Saxafrage and Dodder, each a handful; bruise and boil them in a quart of White-wine, and a pint of Vinegar, until a third part be wasted; then add a handful of Bay-salt, and half a pint of Olive-oil, with half a pound of Honey; and having strained out the thin part as hot as may be, let him drink it fasting: and so continue to do for the space of a Week together, and you will find the Pains will cease.

To take away a Wen in the Neck, or any part of the Horse's Body, without danger.

These Sorrances are generally caused by the assembling of bad Humours to one place, and there contracting into a Tumour. To remove them therefore,

Take the Oil of Bays, Water of Tatar, and Soap-boiler's Lees, mix them well; and being very hot, dip a Cloth therein, and lay it upon the place grieved, continuing often so to do; and the Humours thereby being dispersed, the Swelling will sink and disappear.

The Crownet-scab, what it is, together with the Cure.

This is a troublesome Sorance, being a Scab round the corners of the Hoof, very cankerous and dangerous, and frequently comes by a Horse's running in wet and miery Ground, especially in Winter-time, whereby the Cold has Power to contract the gross and disordered Humours; and is known by the Hairs standing up, the unevenness of the Crownet, the watery Humour that proceeds from thence: Wherefore to cure it,

Take Verdegrease an ounce, Rusty-bacon-fat two ounces, Powder of Hart's horn an ounce; wash the place with Beef-broth, and having made the before-mentioned Materials into an Ointment, anoint the place as hot as may be endured; and continue so to do for a Week together: after which anoint it with Oil of Bays or Rosemary.

To draw out a Thorn or Stump, or any Iron or sharp thing got into the Flesh.

If you cannot come at the cause of this kind of Sorance, so as to draw it out with your Fingers or Pincers, then mollifie the swelling or pat where you conceive it to be; and,

Take of Burgundy-pitch an ounce, and of Black-soap two ounces; stamp the Roots of Water-lillies to the quantity of both the former; and spreading them Plaister-wise, lay them to the place a Night and a Day, and thereby the swelling will not only be sunk or depressed, but the head of the Thorn or Iron will appear to that degree, that it may easily be taken out: after which apply a Plaister of *Diaculum* or *Oxecroteum* to bring away the festered Matter, if any be found there, and so heal it with Green-ointment.

For a Strain in the Coffin-joynt, or Socket of the Hoof.

This Sorrance happens by a sudden short slip, and is not thought of by many Farriers, who are of the Opinion, That under the Hoof there can be no slip nor strain, though the contrary appears; and this is found out by taking up the Foot, and bowing the Hoof from side to side; and on that side you perceive the Horse to be pained in so doing, on that side is the danger. This being perceived,

Take of Beef-brine a Pint, and as much tried Suet; set them on the Fire, and let them consume to a third part; then add Wheat-meal and the Juyce of Ale-hoof or Ground-ivy half a Pound, or so much that they may be made up into the thickness of a Poultis; when having pared the Hoof at the bottom to the quick, spread some of it very hot, and stop it in with Flax: and so renew it every other Day, till you perceive by your Horse's going the Grief removed; and the better to keep it on, you may clap a cross Stick under the Shooe, or cover the whole Foot with a pitched Cloth, or a thick sole of Leather will do the same Office: but observe that during the Cure, you suffer him not to go in wet and dirty Ground.

For a Horse that is perpetually sick or out of order, by retaining a spice of former Surfeits.

Having let your Horse blood take him up into a warm Stable, and prepare two Ounces of Alloes Sucatrina, finely beaten to Powder; rowl them up in Buttur that has been tempered with the Juyce of Rue, and give him the Composition in Pellets as big as Walnuts, in the Morning fasting, having prepared his Body for the reception thereof, by dieting him some Days before with scalded Bran and boiled Barley, refusing on this occasion to give him hard Meats till three days after, suffering his drink to be white-water, and that very warm. Now this white-water is no other than water wherein flower or bran is scattered, or bread finely grated and sifted, *&c.*

For a Cold in the Summer when the Horse is defective in filling, or by too rank feeding.

Take of the Powder of the Root of Scabeous an ounce and a half, of Red-stone-sugar half a pound, and of Olive-oil four ounces; add to these half a pint of Canary, or as much as will dissolve them to a thinness, that the Horse may easily swallow them; and so give them to him luke warm in a drenching horn; and after them a gentle heat, in such manner, that the humours may be stirred and brought away by excrement of sweat, urine, *&c.* And this may be given in case of any extraordinary over-heating, by racing, and the like.

A Through-splint or Screw-pin, what it is, together with the Cure.

This Sorrance happens on both sides of the legs so opposite one on the other as if it riveted it; and from thence takes the denomination of Screw-pin, and is a kind of a sinewy excrescence: Wherefore to remove it, for it is very troublesome to the Horse,

Take the handle of a hammer, or the Blood-stick, and beat it therewith on either side, till you have reduced it to a softness; then having an ounce of the Oil of *Riggrum*, or by some called *Reggium*, anoint it therewith, and bind a cloth about it, and by frequent dressing the Excrescence will sink, and restore the Leg to a flatness; for want of the before-mentioned Oil, take that of *Petrolum*, and it will work much the like effects.

To render a brittle Hoof firm and serviceable.

Take the Juyce of Garlick and Rue, or Herb-a-grace of each four Ounces, Powder of burnt Roach-allom, half a pound, Hog's grease and new Cow-dung, of each a pound; mix them together, and being made into a Poultis over a gentle fire, apply it hot to the Hoof, binding it about it, and stuffing it in under the Shooe; and in often renewing it the Cure will be perfected.

To Cure the Anbury,

Sear it off with a sharp hot Iron, and having Hog's grease and Verdigrease well-tempered together, anoint the Roots therewith at sundry times, &c.

To prevent a Horse's pissing Blood, an excellent Remedy.

Take the Herb called Harts-tongue half a handful, Piony-roots sliced an ounce, the Juyce of Bettony half a pint; boil them in stale beer, and give them him as hot as he is capable to endure, the liquid part only; and so continue to do in the morning fasting for a week together, and the defect will cease.

For a broken Knee, the Remedy.

Take Urine, heat it well, and wash the Sorrance, easing it of the Gravel if any be contracted, by falling of the like; then take Turpentine an ounce, Rosin the like quantity, Hog's Lard two ounces, and the Blades of Leeks a handful, with two ounces of the Powder of burnt Allom; bruise with Leeks, press out the Juyce, and melt it with the rest, to the thickness of a Salve; and spreading it Plaister-wise, apply it to the place grieved, anointing it between whiles with Oil of Coriander-seed or Anniseeds.

For the Dropsie in a Horse.

This Distemper proceeds from the looseness of the flesh, occasioned by moist and unwholesome feeding, whereby flegmatick and watry Humours are ingendred in the Blood, which sweating as it were through the Veins, are retained between the skin and the flesh, or in the spongy flesh, occasioning Tumours and unseemly Swellings. To remove which,

Take two handfuls of Parsley-seeds, the like of Anniseeds and Bay-berries, with one handful of Juniper-berries; bruise them together in a Mortar, and boiling them in Verjuyce, sweetened with brown Sugar; give the Horse to the quantity of a pint to drink, first and last chafing the swelled or tumourous places with your Hand, or hard wisp of Hay; and so continuing them for a Week together, you will find the flesh become firm, and the watry Humours disperse.

To joyn a Sinew that is out.

Cleanse the Wound with the Juyce of Nettles and white Sugar; then take the Ointment of Tobacco, and a Plaister of *Diaculum*; apply them and bind up the Wound very strait.

For a Wound or Hurt in the Tongue, a Remedy.

This Sorrance being occasioned by the Halter or Bitt, in having to do with too hard a hand, To cure it, Take of the Juice of Sallendine half a pint, as much of that of Bugloss; heat them over a gentle fire, adding two ounces of Honey or Roses, and as much Allom; and with them anoint the grieved place, and you will find it quickly heal.

For the Itch in the Tail, or any other part.

This Disorder proceeds from rank Blood, through foul feeding; therefore having bled your Horse well,

Take Wood-ashes a peck, Burdock-roots a handful or two, Man's Urine two gallons, and the like quantity of Water wherein Tobacco stalks have been steeped; boil them up into a Lye, and with it wash the grieved part when it is very hot.

Another excellent Remedy for a Tetter.

Take of the Roots of Elicampane and red Dock, of each a handful; steep them a Week in Urine, adding two handfuls of Bay-salt; boil them in the Urine, till from two quarts it became one, and with it wash the Sorrance, after it has been well rubbed and chafed, that the liquids may the better sink in.

For a Hurt or Wring in the Withers.

This commonly happens by the straitness of the Saddle, or indiscretion of the Rider, and is known by a Swelling or Tumour on the Back, &c. To remedy this,

Trake a handful of Wheat-flower, wet it with half a pint of Whitewine Vinegar, adding three ounces of Honey, and an ounce of the Juyce of Hemlock; mix them well over a gentle Fire, and apply them Plaister-wise to the place grieved: or for want of these,

Take Water wherein Barley and Fennel have been boiled, and wash the places with it as warm as may well be endured.

Worms of any sort in the Body of a Horse, how to kill and remove them.

Take a handful of the tops of Broom, and of Savin the like quantity; bruise them together with as much Feather-sew, then with fresh Butter and Treacle make them up into Balls the bigness of Pigeons Eggs; and when he is fasting in the Morning, give him three of them, and let him fast three Hours after; then give him Oats, but restrain giving him Water till the Evening; and in this doing four or five times, the Worms will be destroyed or evacuated.

An excellent Salve for any sort of Wound, how to make and apply it.

Take of clarified Rosin a pound, the like quantity of Bees-wax, Sheeps suet half a pound, Frankincense and Storax, or each half an ounce, Gumaraback four ounces, Hog's Lard a pound and a half, and of the Juyce of green Tobacco half a pint; dissolve and melt them over a gentle Fire, adding in the melting four ounces of Turpentine, and the like quantity of Red-wine; and so boil them up into the thickness of a Salve, by drawing off the watry part, if any remain, and apply it Plaister-wise on Leather of thick Linnen, in case of any Wound, Sorrance, Bruise, or other Affliction.

An Imposthume, to ripen or cure.

The greater or lesser quantity of Matter relating to the Imposthumation, may be discerned by the heat, being more or less, as also the throbbing and beating; then to break it, apply the Roots of White Lillies and Marshmallows, the Flowers or Roots of Mareblabs; bruise them in the best wise, in a Mortar with Hog's Lard and Wheat-flower, and Poultis-wise apply them to the Swelling; which being thereby brought to a head, Lance it; and having drained the Corruption, apply a Plaster of the before-recited Salve; and by often renewing the Plaister, it will draw the Corruption from all parts, and render the Horse sound and safe.

For any Internal Sickness, another good Remedy, never before published.

If you would have your Horse be mostly exempted from Infirmities, especially such as are Internal; after having let him Blood in due season, or as the Malady require it, and given him, if need require it, an opening of Clyster made of Camomil, Marshmallows, the Flowers of Arch-angel and Comfrey, boiled in Milk or Ale.

Take round *Aristolochia, Gentian*, and the Roots of three leav'd grass, of each two ounces, Race-ginger, and Nutmegs, of each an ounce, the Seeds of Cardamoms, and the Juyce of Hysop, each two ounces, Indian Spikenard two drams, Licorish an ounce, *Diascordium* two ounces, Raisins of the Sun a pond, and twelve blew Figs; boil these in a Bottle of White-wine, till a third part be consumed, and give it the Horse to drink warm.

This is successively give in case of *Surfeit, Feaver, Pestilence, Pains in the Belly* or *Stomach, Internal Bruises, Inflammations of the Liver,* or *Consumption of the Lungs,* and the like.

For the Ach, Weakness or Numbness of the Joynts.

The occasion of this Grievance comes either from an unhappy Strain, or a Contraction of gellied watry Humours, caused by Cold, and too much Moisture: To cure and remove which Aches and Pains, *&c.*

Mix *Acepium* with *Canary,* and when you find it very warm, anoint the place grieved therewith, chafing it in with your hand, or a hot cloth, and in a week's time it will remove the Disorder, especially if it proceed from Cold; but if from a Strain, then take *Aqua vitae* and the flower of Brimstone, with a small quantity of the Oil of Spike; heat them hot, and with a hot cloth anoint the place grieved, bathing and suppling it in, swathing the place about with a Rowler dipped in melted Bees-wax and Hog's Lard.

How to stay any violent Loosness.

This disorder in the Body happens many ways, but especially by raw and unwholesome feeding: To remedy which,

Take of the Juyce of Sloes a pint, half a pint of the Juyce of Sene-green, Bean-flower and Bole-armorick, each two ounces, Allom one ounce; boil them together to the consumption of a third part; and then making that up with Milk give it the Horse fasting; and so do for three days successively.

The Lampas; *what it is, and how to Cure it.*

The *Lampas* is no other than a swelling that proceeds from rank Blood, and appears in proud flesh on the inside of the Lips, the way to desroy it is, when you have gagged the Horse's mouth that he cannot shut it, take an Iron with a flat top, the shape of which you find in the Margin; and heating it red hot, burn away the flesh, and then rub the place with a roasted Onion and Bay-salt, and in a short time the swelling will disappear, and the Horse be better able to eat his meat.

For the Fig in the Foot of a Horse, the Cure.

This Sorrance is a lump of unnatural Flesh that grows upon the Frush of the Heel, in the shape of a Fig, and is frequently caused by the defect of those that undertake to cure a wound or hurt in that part occasioned by a stub or thorn.

Cut away so much of the Hoof that by the means of the Incision there may be a space or difference between the Frush and the Sole; then dip a Spunge in an Oinment made of Verdigrease, Bees-wax and

Hog's Lard, and binding it hard on, it will in three or four times so doing destroy the proud Flesh, and render the Horse's Foot sound and well.

For the Flanks, a Disease so called, the way to remedy it.

This proceeds from a Wrench, Stroak or Pain in the Back which causes a Swelling ,*&c.* And to prepare it for the Remedy, shave away the Hair where you find the Frief. Then make a Charge and apply it; which may be made with success after this manner:
Take *Bole-armoniack, Consolida Majora, Galbanum, Mastick, Per-rosin* and *Apoponax,* of each two Ounces or less, according to the largeness or smallness of the place grieved: bruise them well to Powder, adding of *Sanguis Draconia* and *Sol armoniack,* of each three Drams; Wheat-meal, and the Whites of Eggs, with so much Vinegar as will make them into a Cataplasm or Plaister, commonly called a Charge, and lay it warm to the place grieved, being spread upon a Sheep-skin: And this being renewed four or five times, will effect the Cure.

The Shackle-gall, and its Cure.

This is generally occasioned by the fretting of the Shackle or Fetlock, and sometimes by Ropes with which the Horse is tied: To cure which,
Take a good handful of Plantane, and boil it in Milk to a softness; then take six Ounces of Allom, and two Ounces of Sugar-candy, and put them in, beaten to Powder; then add as much Vinegar as will make a hard Curd come on the top: then the Curd being taken off, wash the place with what remains; and then the Hair being clipped away, anoint the place with Hogs lard, and the Powder of Turmerick; or you may do it with the Ointment of Tobacco, or Honey, Verdigrease and Red-wine, made up into an Ointment. And this likewise is good for a gall'd Back, or a Gall in any part of the Horse's Body.

A Horse that is graveled, how to remedy.

When by the Horse's crimpling and lameness, you perceive he is graveled by travelling in foul Ways, take off his Shooes and search well his Feet, clearing away all the Stones and Sand you can find there: Then, to restore him,

Take Bees wax an Ounce, Deers-suet, Rosin and Board-grease of each an Ounce, and four Ounces of the Juyce of Housleek; make them into an Ointment, and apply it hot on a Wad of Flax, stopping it in.

Of Cataplasms or Poultises

Notwithstanding what has been mentioned, there are divers Cataplasms or Poultices of singular use, and, as the best of this kind, take the subsequent.

For any Swelling, Imposthume, Rankling, Wound or broken Bone, a Cataplasm or Poultis.

Take of new Milk one Pint, crumble into it White-bread, and boil it till it become thick; add then the White of an Egg, and an Ounce of Olive-oil; mingle them well together, and apply them Poultis-wise to the place grieved as hot as conveniently may be endured,

For a Swelling in the Throat, or under the Ears.

Take Neets foot-oil a Pint, of the Leaves of Marrigolds a Handful, Saffron a Dram, and of White-bread four Ounces; boil them together till they become the thickness of a Poultis, and apply them hot to the place grieved.

To draw or break a Boil or Ulcerous Sore, &c.

Take the Flowers of Lady-cups, the Roots of Cuccow-pintle, the Leaves of Burdock, and the Flowers or Roots of Water-lillies, of each half a Handful: boil them in a Quart of Linseed-oil, and mash them into a smallness: which done, apply them Plaister or Poultis-wise, binding them hot to the place grieved, as conveniency directs.

An excellent Poultis to asswage any Pain or superate Tumour.

Take the Leaves of Mallows half a Handful, of Grounsel one Handful; boil them in Running-water till they may be made into a Mash; then add to them a Pint of Cream and two Ounces of crumbled Wheaten-bread, of Mutton-suet half a Pound, Oil of Roses two Ounces, and the Whites of two Eggs; boil them all till they become the thickness of a Poultis, and in that manner apply it to the place grieved, renewing it as you see occasion.

A Poultis to break any Infectious Sore.

Take of Lilly roots two Ounces, Marshmallows and Violet-roots, of each the like quantity; Rye-meal and Linseed oil of each four Ounces; of Barley and Wheat-meal, each an Ounce; to these add two blew Figs; of the Flowers of Camomil half an Ounce; and the Bark of Elder-roots the like quantity; boil them stamped and well bruised till they may be strained into a Pulp; then add again Barrows grease and Oil of Almonds, of each two Ounces, and apply it to the place till you perceive it drawn to a Head; after which you may lance it, and with a Plaister of *Diaculum* draw out the infectious Matter, and heal the grieved part.

To disperse the Flux or Oppression of Blood in any Part.

Take Frankincense, Alloes, Dragons-blood and Bole-armoniack, of each half an Ounce, the Whites of two Eggs, and the Wool of an old Hare; mingle them well in a Pint of *Aqua-vitæ*, and let them boil till they come to a thickness; then Plaister-wise apply them to the place grieved as hot as may be endured.

For an Imposthume or sudden Swelling in any part of the Body.

Take French Barley a Pound, bruise it with the like quantity of Linseed, shread them to a handful of Marshmallows, and seeth them together in a Quart of new Milk till they come to a solid thickness; then, as hot as may be endured, apply them to the Swelling, and so continue to do for four or five days successively.

In the case of the Palsey in the Head (a Disease seldom happening in Horses) apply this Poultis, viz.

Take a large Onion, roast it well, then put to it an Ounce of the Oil of Spike, Olive oil and Lavender flowers of each two Ounces; Marjorum and Winter-savory, of each half a Handful, well shred; boil these, and apply them as a Poultis, hot as may be well endured. And thus much, in brief, of things in this kind.

Rare and New Experiments

As for Balms and Balsoms, seeing they are wonderful useful on sundry occasions, I think it not a miss to speak something more of them and their Use by way of Receipt. And first,

To heal and contract any Wound.

Take of Mummy three ounces, Alloes Epatick half the quantity, Stone-pitch two ounces, Sarcol half an ounce, Gum-arabick and Mastick, of each a quarter of an ounce; add to these half a Pint of *Aqua vitæ*; melt and order hem over a gentle Fire till they become a Balm; and then, as occasion serves, dipping into a Lineament in the contracted Medicament, apply it to the Wound as often as is convenient.

An inward Balm to destroy Worms, and heal Internal Bruises.

Take Oil of Turpentine half a pound, Myrrh, Storax and *Galbanum*, of each an ounce; Cloves and Cinnamon in Powder, of each half an ounce; Deers suet half a pound, and of Amber-grease a dram; make them into a Balm over a gentle Fire, and give the Horse in warm Ale the quantity of a Walnut fasting.

An excellent Balm in case of any Sprain, Internal Bruise, Swelling, Blasting, old Sore or Gun shot.

Take of Turpentine an ounce, *Galbanum* two ounces, Cicatrine, Mastick, Cloves, Galingal, Cinnamon, Nutmegs, Cubeba, of each an ounce; Gum of Jape half an ounce; beat them and well incorporate them, then distil them over a gentle Fire in a Glass; and when the thinnest part is drawn off, the next will be a Red Oil, and that is it which is to be applied with the greatest success, though the first is wonderful effacious and useful on many occasions.

An excellent Balm to be given a Horse inward, in case of a Consumption.

Take of the best Turpentine a pound, Pine Rosin a fourth part, Myrrh, Frankincense and Mastick, of each two ounces; Sarcoco, Mace, Wood of Alloes, of each an ounce, and of Saffron half an Ounce: put them in a Glass Retort in hot Ambers; and after the Water, or the clearer part is drawn off, there will come forth a reddish Oyl, which may be given him four drams in a Morning fasting, suffering him to take it in half a pint of warm Ale.

An excellent Red Water to cure Ulcers.

Make a Lye of Ashwood-ashes, that a gallon of the liquid part may be drawn from it; and add to it a gallon of Tanners Ousey, in which no Leather has been steeped; steep in these two pounds of Madder, and dissolve half a pound of Roach-allom; then let them simper over a soft Fire till a third part be consumed; after which run it through a fine Sieve, and dust into it Bole-armoniack: which done, set it again over the Fire till it become to half the quantity: and so, as you see occasion, you may wash the place grieved with it.

An excellent Water to allay any Internal Heat or Feaverish Indispondency.

Take of Savory, Sorrel, Bugloss, Burrage and Ednive, of each a handful: chop them small, and boil them in two quarts of Running-water, scumming it till half be consumed, and add as much Verjuyce as will make up the first quantity; sweeten the liquid part with brown Sugar, and reserve it for your Use, giving the Horse half a pint at a time fasting.

In case a Horse be troubled with the Stone, an excellent Water.

Take two quarts of new Milk, and of Saxifrage, Parsley, Mint, Fennel, Pellitory of the Wall, Mother of Thyme, green Sage and the Roots of Radishes, each an ounce: bruise the latter, adding two quarts of

White wine; and so, if your conveniency will admit, distil them, if not boil them, and strain out the liquid part, and give him half a pint at a time, having first scraped into it the Powder of a roasted Nutmeg.

For any Disease in the Eyes, another excellent Water, &c.

Take of Maiden-hair and Ground-ivy, a handful of each, Alabaster two ounces, and of the Roots of Wormwood dried and beaten into Powder an ounce: distil or well concoct these, and when you have so done, wash the Eye grieved, by dipping a Feather into the liquid part.

An excellent Purgation for Gravel in the Bladder or Kidneys.

Of Parsley-roots take a handful, white Saxifrage and Ashen-keys, or the bark of the Ash-tree-root, of each an ounce; or Paristone, a Herb so called, half a handful; Eringo-roots sliced two ounces; boil them with half a pint of Coriander-seeds in a gallon and a half of new Ale, and give the Horse a pint of the liquid part to drink, as hot as he can well endure it.

For the Ulceration of the Yard, an excellent Water.

Take Spring-water a gallon, quench in it a hot Iron and Flint-stones very often; infuse into it the Leaves of red Roses, or Rose-cakes, four ounces; Pomegranate-peels, and the flowers of the same, each half an ounce; add of the Juyce of Plantane and Housleek, each half a pint; of Allom and white Copras each half an ounce: boil them over a gentle Fire, and inject the liquid part into the Yard with a Syringe, and it will effect, in often using, the Cure.

An excellent Powder for the Falling-sickness or Falling-evil in a Horse, &c.

Take the Roots of Elder, dry them in an Oven till they may be beaten into a Powder; add the Powder of a roasted Nutmeg and Storax, each an ounce; the Ashes of the Wool of a Fox, half an ounce; and of the Powder of Calamint a dram; mix these well, and give him half an ounce at a time, in a quarter of a pint of Canary.

A Powder for the Ague, which frequently happens, especially to young Horses.

Take the Herb Mercury, Plantane-leaves, *Cardus Benedictus* and Rue, of each half a handful: dry them, that they may be beaten to Powder, and give the Horse an ounce of it in a quart of Ale, wherein two handfuls of Centaury have been boiled, as hot as he can well endure it: and so renew the Dose as you shall find occasion.

To Purge Choler and Flegm, an excellent Remedy.

Take of Turbith an ounce, Ginger, Cinnamon, Mastick, Gallings, and Alloes Epatick, of each half an ounce; Diagredium, Rhubarb and Senna, of each a dram; dry, bruise and make them into a Powder, giving the Horse the whole quantity at two Doses in warm Ale or Milk.

An excellent Bath to allay any Swelling, or such-like Disorder, especially Diseases in the Legs, occasioned by the descending of evil Humours, &c.

Take the Roots and Bark of Pomegranate; the Flowers of Comfrey, and of Acrons, each a handful, Camomil and Fumitory, of each a handful, black Helebore and Hysop the like quantity; boil them with a gallon of Water, and when a third part is consumed, apply with a Woolen cloth the remainder to the place grieved as hot as may be well endured.

A Bath to soften and mollifie the Skin.

Take the Roots of Marshmallows and white Lillies; bruise them with Fenegreek-seeds, Peletory of the Wall, and Violet-leaves, the flowers of Camomil and Melliot, each an ounce, Neats-foot-oil, the Oil of Lilllies and Hog's Lard, each four ounces; add to them all a quart of Water, boil and strain out the liquid part, using it as hot as may be by way of Application, *&c.*

To stanch Blood in any Vein or Artery.

Take Alloes Eparick and *Olibanum*, of each half an ounce, and the Wool of an old Hare; bruise them with the White of an Egg, and spread them on Cotton-wool, binding them to the place, and there suffering them to continue till such time as you find the Blood is turned back, and the Film knit together, which will be within the space of two or three days.

An excellent Medicament to provoke a Horse to Vomit, as also to purge the Belly.

Take Elder root-rind, bruise it small, to the quantity of two ounces, the like quantity of Spurge-lawrel and Turmerick; let them steep in a pint of White-wine a night and a day, and give the liquid part to the Horse very warm.

An excellent Purge, good on sundry occasions.

Take Senna, Coriander-seeds, Alloes, and the Juyce of Savin, of each an ounce; steep and bruise them in a quart of Ale, then give him the liquid part fasting, as hot as may be well endured, for two mornings successively, ordering them well to his Dressing and Diet, that he neither over-feed nor catch Cold.

For a Joynt-sickness.

Take Ants-eggs, together with some of the Ants, a small quantity, add the Keys of an Ash tree, the Roots of Briony, and those of Burdock; boil them in Whey, and with the liquid part anoint the Joynts as hot as may be well endured, binding up and keeping your Horse warm, *&c.*

For an Internal Ulcer.

Take of Bees wax four Ounces, Turpentine the like quantity, Conserve of Red-roses an Ounce, Deer's Suet two Ounces, Storax half an Ounce, Myrrh the like quantity, and Oil of sweet Almonds, as much as will make them into a Balsom, and give it the Horse, an Ounce at a time in a Pint of warm Ale.

An excellent Electuary for a dangerous Cough or ratling Cold.

Take Germander, Horehound, Hysop, Agremony, Bittany, Liverwort and Hearts-tongue, of each a handful, clean stripped and washed; boil them in three Pints of Water till they are very soft, and till the Water be consumed, that they may be mashed into a thickness; then add the Powder of Licorish, Elecampane-roots and Honey, so much as will make it into an Electuary; and by so doing, and giving each Morning the quantity of a Walnut to your Horse fasting, it will wonderfully help him.

Many may be further taken notice on of this kind, but these being the rarest, newest, and the best approved, I hope the Practitioner will have such Satisfaction herein, that he will need no more.

Chapter XXIII

The Symptoms of Diseases, Sorrances, Distempers, Grievances, or the like, in general and particular, how to foresee them, and prevent them; as also to know when they happen, &c.

Having passed over all the material Diseases and Sorrances any ways incident, hurtful or dangerous in the plainest, fastest, and easiest Method; it now remains that I speak something of the Symptoms of Diseases and Distempers in general, that they may be the brieflier comprehended, and afterward give Directions for making Unguents, Salves, Poultices, Clysters, Suppositories, Purging-potions, &c. which ought always to be kept in a readiness for sundry Uses and emergent Occasions, with such reasonable Directions as may add to what has been already treated of: and of these in their order.

The Curious, as I have formerly hinted, have ever had a great regard to the Complexion of a Horse, thereby to draw from thence more than bare Conjectures of the bodily State or Constitution relating to Health or Sickness. And since too much of this kind cannot be well laid down, nor more than sufficiently handled. I shall make these further Observations than what hitherto I have made:

The Strangles are signified by the hanging out, and unseemly colour of the Horse's Tongue, and by the faintness of his breathing.

When a Horse is very thirsty, and seems very little affected towards his Provender, then has he the symptoms of a Fever, or some hot Disease that afflicts the Heart and Liver, or else it may signifie the Putrification of the Lungs: but when he eats largely, and desires not much Water, it denotes a cold Liver, and that the Horse is subject to gross Humours, by reason the Heat cannot concoct the quantity of Nutriment as it ought, and therefore it is not amiss to restrain him from eating altogether so much as he requires, or at least to give it him by degrees, that it may leisurely digest.

If with exceeding greediness he devour his Meat and Drink, then beware he be not troubled with the diseases of the Spleen, or Putrification of the Lungs.

If the Breath of a Horse, without travelling or other force or violence, be found very hot, or so much as is more than usual, it denotes the Symptom of some feverish Disease approaching.

If the left side be much swelled, and there be no apparent cause, then proceeds it from the disorder of the Spleen; and if the Legs on that side be likewise swelled, then it commonly proceeds to the Dropsie.

Drivelling or noisome Water descending or issuing from the Mouth or Nostrils of the Horse, denotes the wet Cough; and if it be gellied, or the like, then it threatens him with the Staggers.

The dullness of the Countenance, lolling of the Ears, and hanging of the Head, are signs of the Megrim, or extraordinary Pains in the Head.

If disorderly Pantings appear on the Breast, Sides, or any part of the Body, then does the Horse labour under some Sickness that afflicts the Heart or Liver.

If the Mouth be foul or furred, and the Tongue look yellowish, then the Lungs are defective, and tending to a Consumption.

The hollowness of the Temples, denotes either the Strangles, or that the Horse is very old.

Shortness of Breath, hanging of the Eye-lids, and beating of the Flanks, denotes a Fever.

A cold Swelling under the Throat, with a ratling in the Head, signifie the approaching of the Glaunders: If about the Tongue-roots small Knobs appear, then it signifies Cold, *&c.*

If the Horse offer to Couch, and be faint in so doing, as not throughly able to bring up what he offers at, then it is occasioned by the swelling or rising of the Lungs, or oppressive Flegm settled there, which obstructs the Lungs in the performance of their Office.

The staring up of the Hair, and hardness of the Skin, with dejected Looks, and lankness of the Belly, denote the Horse foundered in the Body, and sometimes the Wind-cholick, or Stone is signified thereby, as also the Yellows; which are all dangerous Distempers in a Horse.

If the Skin stick to the Ribs, so that it cannot be well raised, then the Horse is troubles with that Infirmity which we commonly call Hide-bound.

An uneven stiffness in going, denotes some strain, wrench, cold swelling in the Joynts, or foundering in the Feet, *&c.*

If a Horse have a spongy Wart full of Blood, it is an Anbury; if a knotty Ulcer creeping along the Vein, it is a Farcy; if Scabby or Ulcerous upon the Body, and about the Neck, it proceeds from the Mange; if it singly spread abroad, and that but in one place, then is it held to be the Canker.

The Botts, or such like Insects in the Paunch or Belly of a Horse, you may know by the Horse's endeavouring to strike thereat with his Feet, his lying down and wallowing himself, and his often turning his Head back and looking upon his Sides.

If the Horse be over-covetous to lie down on the right side, it signifies corrupt Blood settled in the Cavas of the Liver, and occasions extraordinary Heat, which by the pressure of the Liver is augmented.

A Horse's spreading, when laid down, generally denotes the approach of the Dropsie, and his often groaning, the Cholick, or the Heart's being oppressed with bad Blood: And thus of other Signs and Symptoms, most, or the most part whereof, I have in this Chapter; and what has been before mentioned, is effectually discussed according to the best Experimental Observations that have been made. From whence I shall proceed to give Directions or sundry choice Ointments and Salves, highly necessary to be kept in store, and used on sundry occasions, as necessity requires; and other matters altogether as material.

Chapter XXIV

Directions for Making and Preparing Oyntments, Oyls, Salves, Waters, Purgations, Poultises, Charges, Supplements, Pills, Powders, &c. singular good in case of any Distemper or Sorrance: Many of them never before publick.

Although I have mentioned many famous Oyntments and Salves in the course of the Cures, yet some there are which may indifferently serve for most Sorrances and Griefs of any kind; and these I shall chiefly name, and direct how to make them, because they may be gotten in a readiness, and thereby the Party not be to seek them, nor his Ingredients, when the urgency of the Horse's Distemper or Grief requires the Application.

An Oyntment to search any Wound or Ulcerated Sore, or anything of the like nature.

Take of Bees-wax four ounces, Turpentine the like quantity, the Juyce of Spurge-lawrel two ounces, Deers-suet half a pint, Verdigrease an ounce, Allom calcined two ounces, and Hogs-lard as much as will make it into an Oyntment over a gentle Fire. This, by often using, will not only search the Wound, and discover dead, proud, or putrified Flesh, but cleanse it, and cause it to heal, restoring good Flesh, and rendring it easie to cure.

An Oyntment excellent good in case of Botches, Boyls, Scabs, or the like Sorrances.

Take the Juyce of green Tobacco half a pint, or Deers-suet a pound, the Powder of Dandelyon-roots two ounces, as much of Soap-maker's Ashes, and half a pint of the Lees of Wine: make them up into an Oyntment with half a pint of Olive-oyl, and an ounce of the Oyl of *Petrolum*.

To skin any Wound, an excellent Oyntment.

Take of Dogs-grease two ounces, half the quantity of black Soap, of the Powder of calcined Roach-allom two drams, the Juyce of Mug-wort an ounce: make them into an Ointment over a gentle Fire; and when the Wound begins to fill with flesh, anoint it over.

To mollifie and asswage any Swellings, an excellent Ointment.

Take of Nu toil a quarter of a pint, Neatsfoot-oil half a pint, and Linseed-oil the lke quantity: add to these the Juyce of Plantane a quarter of a pint, and four ounces of the Oil of Earthworm: boil them over a gentle Fire, to a convenient thickness, and apply the Ointment to the place grieved as warm as may be well suffered, and chafe it in with your warm hand when it begins to cool.

An excellent Ointment to cool and allay any Inflammation.

Take of the Oil of Marshmallows half a pint, the Juice of Mandrake roots two ounces, Dogs fat four ounces, and fresh Butter a quarter of a pound: make them into an Ointment over a gentle Fire, and, as occasion serves, apply them to the place grieved till you find the extraordinary heat abate.

An excellent Ointment, or rather Balsam to be inwardly given a Horse for Obstructions, Bruises, and other Ailments and Grievances.

Take of refined and rarified Turpentine two ounces, Stage-suet the like quantity, Amber-grease two scruples, *Olibanum* an ounce, Oil of Roses two ounces, and a dram of the Oil of Amber: heat these gently till they incorporate, and then take a small quantity, and with fine Flower make a Paste to the bigness of a Walnut, and give it the Horse to swallow; giving him after it half a pint of warm Mallaga.

To make a green Ointment, proved by Experience more effectual than what has formerly been published.

Take the Juice of Sage two ounces, as much of that of Rue, an ounce of Verdigrease, and of *Aqua-vitæ* half a quartern: mix these over a gentle Fire, and add of the Powder of Elicampane-roots an ounce, with the Powder of white Copras calcined half an ounce: make them into an Ointment with Olive-oil, and half an ounce of the Oil of Turpentine.

An excellent Remedy for the Staggers of any Pain that suddenly takes a Horse, sometimes to the loss of his Life.

Take of the Fat of the Guts of a Capon two ounces, Oil of sweet Almonds two ounces, of Olive-oil half a pint; incorporate them well over a gentle Fire, then drop into them the Chymical Oil of Nutmeg a dram, and the like quantity of that of Spikenard and Bay-berries; and having incorporated them farther into an Ointment, when you perceive your Horse afflicted, dip a Feather in the Ointment, and thrust it up the Nostrils of the Horse, and anoint them as high as may be; then burn under his Nose Storax on a Chafing dish of Charcole, placing a Tunnel so over it, that the smoke may ascend into the Nostrils only. This is likewise good for any cold Rheum that afflicts the Head, and will bring away the superfluous humour.

An excellent Salve for any Wound.

Take Hogs-lard half a pound, Bees-wax a pound, Stone-pitch six ounces, unslacked Lime beaten into Powder an ounce, the Powder of dried Fox-lungs an ounce; make them into a Salve with two ounces of Turpentine, and apply it Plaisterwise to any Gangrene, ulcerous Sore, Botch, Strain, Slip, Spavin, after is being opened, or other Sorrance, and by due application it will answer your expectation.

An approved Salve, to draw any Stub, Thorn, Splinter of Bone or Wood out of the Flesh.

Take of Burgindia pitch four ounces, the like quantity of Per-rosin, Nut-oil two ounces, and the like quantity of Linseed-oil, and an ounce of the Juyce of Hemlock: make them into the thickness of a Salve, and apply it Plaister-wise to the grieved part, till you find the head of the offensive matter; and then with your Instrument dilate the flesh, and draw it out.

To fill a Wound, Ulcer or the like with good flesh, a Plaister.

Take Mutton-suet half a Pound, the Juyce of Baum a quarter of a pint, the Ointment of Marshmallows and Groundsel of each two ounces; burnt Allom in Powder two oucnes; Rosin half a pound, and Bees-wax as much as will make it up into a Salve.

A Salve to draw Corruption from the bottom of any Wound, or to draw a Swelling, or any such Grievance to a head.

Take Turpentine half a pound, Linseed-oil half a pint, Chalk beaten to Powder an ounce, the Juyce of Orpin half a quartern, *Galba-nuth* two ounces, and Oil of Vitriol a dram; make them into a Salve, and apply them to the place grieved, and in so doing you will find your expectation answered.

To ripen any Tumour or asswage any Swelling not abounding with extraordinary Humours.

Take of the fat of an Urchin or Hedge-hog four ounces, Tarr two ounces, old Cheese well beaten in a Mortar four ounces, the Juyce of

Garlick half a quarter of a pint, Bees-wax six ounces, and Stone-pith six ounces; make them into a Salve, and apply the Plaister on a piece of Sheeps-leather.

An excellent Poultis for a Tumour or Swelling.

Take Linseed-oil half a pint, the Whites of six Eggs, Bole-armoniack two ounces, Groundsel and Smallage of each a handful well bruised in a Mortar, Celendine and Comfrey the like quantity so ordered; fry them together, and lay them on as hot as may be. This either allays the Swelling, if only fleshy, occasioned by a Stroke or Saddle-pinching, or brings it to a head, in case it proceeds from Humours gathering or contracting in one place.

A Charge to ease a Pain in the Back, or for any Sprain.

Take new Cow-dung four ounces, the Roots of Burdock two ounces, washed and sliced, Borage and Bugloss, of each a handful, Oil of Bays six ounces; bruise them well together, and heating them over the Fire, suffer them to be as hot as may be well endured, and apply them as a Poultis.

To mollifie any Chap or rough Sore.

Take Comfrey, the Roots of Scabeous, the Leaves of Plantain; boil them in Olive-oil, being first well bruised to a softness, then add Neats-foot-oil, half the quantity of the Olive-oil, and then strain off the liquid part, and with it anoint the place grieved.

For the Eyes of a Horse afflicted by any means, an approved Water to cure or ease them.

Take of the Juice of Pimpernel and Eye-bright, of each a like quantity, both consisting of half a pint; add to them the Powder of *Lapis Calaminaris*, quenched in White-wine, an ounce, and as much of the Powder of burnt Allom, two drams of the Calcine of Crabs-eyes, and as much of the powder'd Pith of Oysters; dip a feather in them, well mingled by stirring, and rub it so dipped into the Eyes.

A Water to wash the Mouth in case of any Sorrance or Defect.

Take Spring water a pottle, Roach-allom a pound, and English Honey the like quantity; dissolve them into the Water over a gentle

Fire, and add half a Pint of the Juyce of Hysop, and the like quantity of that of Celendine or Vervine; boil them to the consumption of a third part, and with the Water wash the Horse's Mouth as you see occasion.

A Pill good for any Internal Disorder, &c.

Take of Alloes Epatick half an ounce, Powder of Rhubarb the like quantity, the Juyce of Water-cresses half a quarter of a pint, and the Berries of Juniper dried and beaten into Powder an ounce; make these with the Oil of Myrtle, into Pills as big as Hazle-nuts, and give him four at a time successively in warm Ale or new Milk every Morning.

A Supplement exceeding good for any Strain or Grief in the Sinews, &c.

Take Bacon Lard half a pound, the Oil or Ointment of Smallage two ounces, black Snails a handful, the Powder of Mastick two ounces; bruise and incorporate them, and so apply them to the place grieved.

A Vomit for a Horse that has a queesie Stomach, thereby to render him a good Appetite.

Take Spurge-lawrel a handful, Briony-root an ounce; boil them in a quart or three pints of Water, then strain out the liquid part, and having sweetened it with Sugar-candy, give it him hot, and tend him that he not catch cold.

To purge Melancholy.

Take Scamony a dram, the Juyce or Seeds of black Helebore two ounces; dissolve the former in, and mingle the latter with a pint of warm Ale, and give it him to drink fasting.

To purge Phlegm.

Take of the Juyce of Ivy-leaves or berries half [...] or the Decoction of them being very strong, add the Grains of Cochneal two ounces, the Roots of Fern washed and sliced two ounces, Colocinthius an ounce; make of these a drench with White-wine; give it the Horse warm, and keep him two Hours after fasting.

Chapter XXV

Clysters, how to make them, and on what occasion they ought to be applied in order to their effectual working and bringing away bad Humours.

Since Clysters, in case of Internal Distempers, are very necessary, I cannot omit them, but proceed, amongst other things, to speak of such as cannot but be useful, and especially those that are capable or purging the several Humours, or at least ways to cause an evacuation of those Crudities, they have contracted in the Bowels. Andin this case, if your Horse is very laxative, which frequently is occasioned by flegmatick Humours,

Take of the Juyce of Peletory of the Wall a quarter of a pint; add to it Verjuyce and Olive-oil, of each half a pint, and of stale Beer a quart; boil them together to the consumption of a sixth part, and being warm, put them into your Glister-bag, and force them up the Body of the Horse, and by binding down his Tail, suffer them to remain there, if possible, for the space of an Hour, and give him thereupon Water wherein Scabeous has been concocted.

In case of any Pestilential Disease, occasioned by a cholerick or fiery Humour.

Take of the Seeds of *Coloquintida*, cleared from the husks half an ounce, the Juyce of Centaury and Wormwood of each an ounce, *Castoreum* half the like quantity, Juyce of Wood-sorrel two ounces, and half a pint of Olive-oil; concoct them in two quarts of Water a little sweetened with brown Sugar, and force it into the Horse's Body, using him as before is mentioned.

For any Internal Distemper proceeding from Melancholy.

Take Anniseeds, and the Seeds of Mallows beaten to Powder of each an ounce; boil 'em with a small quantity of Savin in a quart of Whey or Scim-milk; then add a quarter of a poundof fresh Butter, and so having well strained out the liquid part, give it him in his Fundament luke warm.

For any Distemper Internal, occasioned by sanguine corrupt Blood, or watry Humours, by means of bad Concoction or Obstruction, &c.

Take of the leaves and roots of Mashmallows a handful, Violet-leaves double the quantity, Linseed and Coriander-seeds, of each a handful, White lilly-roots an ounce, the Juyce of Senne the like quantity with the latter; boil them in two quarts of Water to the consumption of a third part; then add Olive-oil a pint, and give it to him warm Clyster-wise.

For Sickness in general an approved Clyster.

Take of the Oil of Dill and Camomil, of each an ounce, the Oil of Caffa half an ounce, the Juyce of Violet-leaves two ounces; then having concocted a good quantity of Mallows in two quarts of Water, strain the liquid part, and put the fore-mentioned Ingredients therein, and administer them Blood-warm. This in all violent Diseases especially, is singular good.

In case of Restringency or hard Binding.

Take the Juyce of Fumitory a quarter of a pint, the Syrup of Roses two ounces, and as much of the Oil of Bays, Neats-foot-oil half a pint, and of the Juyce of Mulberries two ounces; add to these a pint of new Milk, and force them up the Horse's Fundament very warm, and so upon other the like occasions.

Observe in giving of Clysters, that the quantity must in all probability be reduced or augmented according to the quality or temperature of the Horse. Now note, That if the Horse be never so large, and in good case, three quarts is an extraordinary dose, and one quart is an indifferent one; so that I leave it to the direction of the Practitioner to regulate the liquid part as he sees convenient, least by over-charging the Horse's Bowels, it burst forth before it has opportunity to work as it ought, as by being under-charged it wants of its force to stir Humours, and cause such an evacuation as is required. And so submitting this to the direction or judgment of those that shall make experiment, I proceed to other matters altogether as necessary to be understood, both as what relates to the preservation of Health, and remedy of Sickness.

Chapter XXVI

Cordials, Cordial-Powders, Drinks and Drenches, Purgations and Suppositories, wonderfully conducing to the Health and Strength of a Horse.

Diapente, *an excellent Powder in case of any Cold or Pestilential Disease. To make it,*

Take *Gentia Baccalani*, Round *Aristolochia*, Myrrh, and the Powder of Storax, of each an ounce; bruise them distinctly, and pass the Powder through a fine Sieve, and when you see occasion, give the Horse from two to four drams in Muscadel, or other sweet Wine, as warm as may be, and keep him from catching cold: or for want of Wine, give it him in strong Ale, but Wine is better.

An excellent Cordial-ball to be given in case of any Internal Distemper, and especially to prevent the Consumption or wasting any part, &c.

Take dried Foxs-lungs an ounce, Methridate two ounces, Powder of Licorish and that of the Seeds of Coriander, of each an ounce; the Powder of Cinamon and Pomegranate-seeds, of each an ounce, Spruge-beer, or the Lees of Claret half a pint: Thicken it with Allom-flower till the whole mass be infused therein, and may be made up in Balls as big as Walnuts; and then give him one at a time, as you see occasion, sending after it a hot Drench of Ale or new Milk: This and the former being put into a Glass or Gally-pot, and close chopped up, will keep in a dry place a Twelvemonth.

An excellent Drench to Cure any Internal Distemper proceeding from any of the four Humours of the Body, &c. *especially such as are Pestilential.*

Having let your Horse blood, if necessity, by the Symptoms I have named in the foregoing part of the Book, requires it,

Take a handful of Balm, and as much Worm-wood; dry them till they may be rubbed into a Powder; mix with the Powder the Juyce of Rue four ounces, the Powder of grated Nutmeg, well dried, an ounce, four grains of the Oil of Amber, and two of Bezora-stone; dissolve these in a quart of Ale, and give the Drench to the Horse as hot as is convenient, and let him stand fasting two Hours in a dry Stable after them.

For want of what is before mention, Take Anniseeds, and the Seeds of Cardamums, of each an ounce, Bay-berries and Fenegreek-seed the like quanity, the sifted Powder of the Roots of Elicampane two ounces, Olive-oil half a pint, and a pint of new Milk; mix them well together, and sweeten the liquid part, whilst it is seething over a fire, with white Sugar, and give it him as warm as is convenient, ordering as before.

Suppositories, and their Use.

This word speaks the intention of the thing, which is no other than preparing before-hand the Horse's Body to receive a Purge or Clyster. Now for the former,

Take Water wherein Wood-sorrel and Maiden-hair, a Herb so called, has been boiled, and give it him to drink with his Provender a day before, or Water wherein Deal Saw-dust has been boiled or steeped; the like you may do with the Leaves of Bay or Holley, or any thing that may be in a readiness for the Purgation to work on.

The other is to be put up into the Horse's Fundament, after he is well raked to prepare him for a Clyster; as a large Candle, a roasted Onion, Garlick and Rue, bruised and made up into a Ball with white Flower, or a Wash-ball; and these must be suted according to the Humour predominant. And in this case, some hold what I shall name to be authentick:

Take, say they, for Choler, bruised Stave-acre and Savin, made in to a Ball as big as a Tennis-ball, with Honey and Bean flower: If for Flegm, Castle-soap a piece as big as a two penny Wash-ball, rowled in Powder of Ginger: If for Melancholy, a red Onion roasted and stuck with Cloves: And, lastly, for Sanguine, or over-flowing of bad Humours, proceeding from the naughtiness of the Blood, make a Suppository of Honey, Bole-armonick, sweet Butter and bruised Spermint; make them up (with Oat-meal small, ground and sifted) into a paste: And to keep any of these in, which must be done for the space of an hour, you must bind down the Tail, and gently Trot your Horse, and when you perceive him sweat, bring him into the Stable, and drawing forth what remains undissolved in his Fundament, give him the Clister you have prepared, and expect the success.

Chapter XXVII

Perfumes, Baths and Purgations, what they are, and to what end they serve; with the manner how, and under what Considerations they ought to be applied.

Perfumes are much available to remove bad scents, or noisome vapours from the Head and Stomach of a Horse, and being moderately and seasonably applied, greatly contribute to his health: and in this case if the Horse be afflicted with cold Diseases, or those that proceed from cold raw Humours,

Take Storax and Benjamine of each half an ounce, *Olibanum* and Frankincense of each a quarter of an ounce, Oil of *Petrolum* an ounce, bruise the Drugs, and make them up in small Balls as big as a pea; burn them upon a Chafing-dish of Coles, so that the smoak may affect the Mouth and Nostrils of the Horse, that he may draw it in with his Breath, and to make him the better so to do, put a gag in his Mouth that it may stand open.

If the Horse be troubled with giddiness in his Head, which is known by the dullness of the Eyes; burn Feathers under his Nose, shavins of Leather or Camel's Hair, or the Seeds of Fennel and Anniseeds. The Roots of Horse-radish or Rhubarb will have the same effect, or any thing of strong scent, whether Gums, Oils, Roots, Herbs, or other things of the like nature and quality.

In case of Bathing a Horse, commonly called the Horse-bathe, it is no more than the concocting divers Herbs that are cooling and supple, with the liquid part wash and supple the place grieved, and render a limberness of the Joynts, or remove Dirt, or any the like Disorder or Grievance, and may many times upon a Journey be used with success to restore or refresh a tired Horse, being applied warm to his Limbs; and the best of this kind take as followeth:

Take Sorrel, Mallows, Groundsel, Camomil, Sparage, Sow-thistle, Comfrey and Scabeous, Endive, Bugloss and Feather-few; boil them in two gallons of Running-water, and when you find they are sufficiently seethed, strain out the liquid part: or if you find your Horse much given to coldness, you may boil them in Chamber lye.

Directions for Purging, according to the estate and condition of your Horse.

If you perceive your Horse of a strong Constitution, not impaired by disease or want, then you may give him strong provocative, that by a conquering quality may prevail against the distemper; but if he be weak and infeebled, then Nature being unable to keep her station, and the potion being strong, great injury may happen, by reason of the fixation or settlement of the gross humours and crudities; wherefore the sudden and violent operation not being capable of removing them, will prey upon, and evacuate those good humours that should strengthen and support the Body: And therefore in case of weakness, a weaker potion must be given, that by long continuance and easie working, it may by degrees loosen and attract what is offensive. And now for a gentle Purge:

Take two ounces of Turnsole, and half an ounce of the Powder of Alloes; dissolve them in a quart of Ale, wherein half a dozen of Lawrel-leaves hath been concocted; give it him warm, and attend the working of it; giving him, the better to hearten him, a Toast dipped in Canary, about half an Hour after; but if the Horse be strong, that he will dispence with a violent Purge; then

Take of *Colloquintids* three ounces, Rhubarb in Powder an ounce, Scamony two grains; dissolve them in half a pint of the Juyce of Hysop, and the like quantity of Canary; give them the Horse very warm, and keep him stirring in a warm Stable two Hours at least, without giving him any thing; and when by a conquering quality they prevail over the Distemper, give him a warm mash, but no hard meat till six Hours after, by which time it will have done working; and in case of cold Distempers you may somewhat augment the Dose. And now as to the Humours, if we take them distinctly, Alloes and Caffa purge Melancholy, *Colloquintida* Flegm, Rhubarb Sanguine, and Scamony Choler. Nor are these the only things that do it, but there are divers others; yet seeing I have largely directed on that occasion, I shall wave them in this Chapter.

Chapter XVIII

Costicks, Corrosives, and Rowelling; what they are, for what cause, and in what manner to be applied.

As for Costicks and Corrosives, they are potable Cauterizings or Burnings, with Oil, Water, or Mineral, Chymically prepared, and sometimes contracted by making a Wound, or breaking the Skin, with Roots and Herbs participating of a fiery Nature, and these are used in eating away dead Flesh, boney or spongy Excrescences; and in case of the Farcy, Mangy, Ring-worm, or the like loathsome dangerous Distempers: and the chief these are,

Aqua-fortis, Aqua-regis, Vitriol, Oil of Tartar, Quick-lime, Oil of Spike, Arsnick, or *Resalgar, Aegyptiacum, Crocus-martis, Mercury* Sublimate, Copras, Verdigrease, Allom, Recordal, and of Roots and Flowers:

Burdock-roots, Horse-radish-roots, the Roots of white Lillies, Garlick, Onions, Cuccow-pintle, Featherfew, Briony; the Leaves of Coleworts, Celendine, the great Southern-wood, Butter-flowers, Ground-ivy; the Flowers of Mare-blabs, Senna, Saxifrage water, Lillies, Holi-hawks, Scabeous, Rue, Bears-foot or Hellebore, *&c.* The Application of which, I leave to the discretion of the Practitioner, to make it as he sees convenient, or consistent with the nature of the Distemper, and so proceed to speak something of Rowelling, another necessary thing to be known.

Rowelling, what it is, and how to be performed.

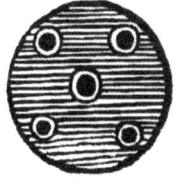

Rowelling is no other than making an Incision in the Flesh, or rather Skin, by taking it up with a Bodkin or other Instrument, so that a ring of Hair, Cord or Leather may be fastned in it; anoint it with Grease, Oil, or other things; thereby to keep the Skin from closing upon it, that in the nature of an Issue the corrupt Humour may be voided; and this is commonly fixed in the Breast or Shoulder of a Horse, for a strain, slip or swelling. And for your better Instruction, observe as thus:

Take a Pen-knife, or other sharp Knife, and slit the Skin right down for the length of an Inch, or more, gently raising it with a flat

Stick, and so do on the other side, about half a Finger's breadth distant; so that the Skin being parted from the Flesh, you may run your Finger quite through, and so put your Rowel in well anointed, and turn it about every other day, freshly anointing it. And though this be the general way of Rowelling, yet the French make a slit or hole only, and hollow the Skin from the Flesh, in circle like a Crown piece; and then cutting a piece of stiff un-allom'd Leather to the bigness, with one indifferent big hole in the middle, and four smaller ones, as the Figure in the Margent; then anoint it with Lard or Oil of Marshmallows, and put it in so that the Orifice may be left open for the Corruption to issue out through a short quill which they fasten therein, running likewise a Needle and Thread through the four small holes to prevent its turning about, or wearing downwards; and so by raising up the Skin, and blowing the wind into the spongy Flesh, cause the Putrefaction to

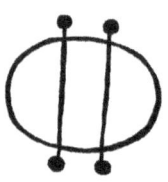

gather and descent, that so by the evacuation of the Humours the Grievance may find a cure; and that it may lie the closer, draw two strong Stitches overthwart, as you perceive the Figure, and let it continue till you perceive Humours well evacuated, and the cutting the Stitches, take it out. And thus I have discussed the material things related to this Science. From whence I proceed to other things highly necessary to be known, as in the Second Part of this Book your discretion may direct you, *&c.*

The Gentleman's New Jockey:
or,
Farrier's Approved Guide

Part II

Containing many rare Experiments relating to Horses and Horsemanship, &c. in such Exactness and Variety as has not hitherto been made publick to the World.

The Manner of Breaking a Horse the best way, and perfecting him in his Paces, &c. and preserving him from Danger, &c.

Having thus far advanced in this great Undertaking, as I well hope with success, I now think it highly necessary, for the better Encouragement of the Reader, to add such rare Secrets, and other Matters, as I doubt not but will confirm him in the sincere endeavor of use to render him in the Practice both Profit and Pleasure. Having back'd your Horse in the latter end of the fourth, or beginning of the fifth Year, and by gentle means rendred him easie and compliable, which by rating, beating, and haring, cannot be so well brought to Perfection, as by cherishing and encouraging him; though sometimes it must be with gentle correction, that he may be the better brought to understand himself. Then the next thing materially to be considered, is to what use you design him that to the same end you may reduce him to pace, amble, trot, gallop, or the like, as either of them may best concur to his advantage. Wherefore, that none might plead ignorance herein, I shall treat of them in their order briefly and effectually:

When you mount a Horse with the intention before-mentioned, you must, if he be an unmanaged Horse, observe chiefly the correction or encouragement of the Voice, Bridle, Switch, Spur, Calves of the Legs, Stirrups, and the Ground; all of these being properly used in the true Management of a Horse; but being to be observed, as the discreet Rider shall see occasion for the best advantage, it would be impossible for me to set down Directions of this kind; though in general I must say, that they must be done with discretion, and in season; or else instead of redounding to the facilitating of Management, they may turn to its disadvantage in rendring the Horse restiff, or hardening him in the Vice of going backward. And now as the true Amble is the justest measure a Horse can take to ground him in, and render him sensible of all other motions, I proceed in the first place to give Directions how, and by what means he must be brought to undertake it with ease and chearfulness, for if the contrary happen, he will hardly be brought to the perfection you wish.

Observe when your Horse is well broke, that he will patiently endure the Bitt and the Rider; that you take him into plough'd Land, not to deep nor cloggy, by reason of Clay or lying low; nor must it be done in wet weather, nor suddenly after a shower; and there rack him a good going pace, till be begins to be a little warm, and feels his Legs well: And so do often Morning and Evening, straining and forcing him every now and then beyond that pace, when as you will in a short time perceive him to fall into a kind of an Amble, but so shuffling, that it is not fit for him long to continue it in this manner, least by custom it becomes habitual; though in ancient times this was the way both held in practiced, whereby the Horses were brought to the motion of Ambling: But since Time and Experience has furnished us with better, it would be Ingratitude not to imbrace them; and therefore, for Brevity sake, passing over the various Discourses and Opinions of some that hold it best for a Horse to fall out of a Gallop into an Amble; of others that would oblige him to it by loading him with excessive weights; and some again, and those not a few, that would have him brought to it by hand between two rails; others there are, and they more erroneous than any I have named, who would have the Horse loaded with Shooes of extraordinary weight to force him thereto, which by often using he must consequently be disabled to perform any motion as he ought:

Part II: Experiments   127

And indeed Opinions in this case are so many and various, that it would be endless in a manner to trace them; many of which, though gilded with seeming Possibility, I willingly omit, and proceed to that which my Experience tells me cannot fail, if rightly managed: Directions in performance of which, take as followeth:

Having trained your Horse, by the help of the plough'd Land, or other gentle means, upon a considerable Rack-pace, to fall into an Amble, though he presently leave it again, and fall to any other motion: Then take strong new Lines made of Hemp or Flax, about the thickness of a Jack-line, well shrunk and dried, that they may neither reach nor shrink in using; let them have Nooses wrought in them at either end of these Lines; there must be two in number of so even a length, that the least disproportion imaginable must not be found in them, their lengths not by any means exceeding thirty seven inches, or being under thirty five; this done, make a soft pare of Hose or Shackles of Girthweb, lines for that purpose, with Wool, Cotton, or soft Linnen; fasten them about the Fetlock-places of the fore and hinder Legs, or places in which Horses are cross-barred to prevent their leaping, and fasten them by Aglet-holes with Leather tabs of equal length, and let the end of a Tab so fastned come from them about the length of eight or nine inches, punched full of holes, to receive the tongue of a Buckle fit for each of them; then having buckled on your Cords even on both sides, to the distance, or as I may term it, the reach of the Horse, when he stands even with his Feet, and proportionably upright, so that by no means they may slip or reach; then take Girthweb, as much as will reach over his Back, and come down to reach the Cords, on both sides, exactly in the middle, wrap it round the Cords, but so as it may not strain or lift them up, nor suffer them to sink lower than their just evenness, and then fasten them either by running a Tag through Aglet-holes made for the same purpose, or by Pack-thread with a Pack-needle; and this is called a Tramel; being one of the best and exactest for this purpose that can be made.

Your Horse being accouter'd in the manner before mentioned, on a plain firm Ground, not over-stoney, move him at first gently, that he may have an opportunity to feel the Tramel, the better to avoid twitching; and so by degrees let him fall into his Amble by moderate managing the Bridle, and holding your Switch between his Ears that he wreth nor wry not his Neck, cherishing him with your Voice, and giving him, when he has done well, some pleasant morsel; but by no means put him to it beyond his ability, least it become tedious to him, and create in him an irksomeness and dislike; and in so using him several Days, that is, every other Day in the Week; if your Horse be sure and good footed, he will take pleasure in it, and perform it without your forcing him to it. But by the way observe, that if your Horse be of a long reach, and when he first undertakes it is consequently subject to twitches, then you may without offence, when he is first put to it, give him a little more liberty, least you subject him to those twitches by sometimes forgetting himself when the Tramel is off, which by the Ignorant, will be taken for the String-halt; but being well used and brought to the feeling of them, you must come to a perfect length, least having more liberty at one time than at another, he proves a Shufflier, or varies in his Amble.

Having brought your Horse pretty well to Amble by the Tramel, insomuch that he does it not confusedly, or against his will, either by hand or backing, you may take him into uneven Ground, and try him in such way as you imagine you may have occasion to ride him in, never standing to pick or chuse your Ground, but rise him as far as is safe, according as your fancy leads you, yet not without the Tramel fixed on one side, which you may shift as you see occasion; and when you find him perfect, you may ride him without, only carry the Tramel with you, in case he make a default; the best way to bring him back to that default, and make him sensible that he has committed an error, which can be him, the time will come, and that within three Months at farthest, that he will be perfect in his Exercise.

There is to be considered next to what has been spoken, the Trot, and that consists in two Parts, *viz.* the lofty and swift Trot; and although to the latter a good Amble is naturally given, yet the former requires deliberation and judgment. And since Experience has afforded a Rule, I think not convenient to conceal it; which, as the best Experiment, I deliver as followeth:

In this case a Horse used to the great Saddle, is the fittingest to be undertaken, one you design for that or the like purpose. Wherefore having chosen your Horse, fasten on him the War-saddle, girting it indifferently strait; then put on a Curb-bridle and fastning to the Chain thereof a Leather-strap, bring it by a Buckle to bear on the Breast-plate or Girths under the Horse's Legs, or which of them you shall perceive most convenient, till you bring the Neck of the Horse to bend or bear Arch-wise to the Neck of a Swain: which done, mount your Horse, and move him a Racking-pace till you find him begin to take up his Legs round and clever, moving his Knees in a manner to his Breast, or bringing the upper and lower spaces between the Joynts to an equal bearing; and so continue to exercise him till you find him move in that manner according to your expectation; and then you may try him without the Straps, and by that means perfect him as you see occasion . Now some there are, that upon the first Backing use to clog their Horses with broad and heavy Shooes; but these I disapprove, and chiefly for two Reasons:

First, Because they greatly endamage the Hoof, and not so only, but by their weight strain and disorder the Sinews, that a Horse must of necessity be obliged to take up his Legs with pain, which will afterward subject him to stiffness and stumbling.

Secondly, They will endanger the Sinews by bruising and cutting, and thereby cause the Bone-spavin, Screw-pin, Ring-bone or Quitter-bone; all which, for the most part, proceed from the like ill Management, and defect of judgment in Managing, &c. But if you would perfect him to the great Saddle, consider farther, that it is highly necessary to manage him at the Ring, that he may be capable of stopping and turning upon the least Motion; and this Ring must be upon soft sandy Ground, where having gently circled him till you have trod out a Ring between forty and fifty Paces in compass, oblige him to trace it, resting him on the right, and cherishing him on the left, stopping him likewise upon the latter, as you see occasion, sometimes in the midst of the Circle, and at other times where he began, shifting likewise, as you see convenient, your hand, and mending your pace, obliging him upon a full stop to retire a pace or two backward; and so from a Trot to a Gallop, ever observing, that Galloping to the right, he leads with his left Foot, and so consequently Galloping to the left, that he leads with the right, and in so doing he will perform it with ease and delight; when, on the contrary, if he carries both his Feet even, leading with neither, he must do it with pain, and be apt sometimes to strike; and this is best in case of a full speed upon a strait Course, and in so using convenient Girts and Furniture, and Keeping a steady Hand, you may stop him upon a full Career, and oblige him suddenly to retire, if any imminent Danger be apparent, or sometimes for your own Pleasure; and by such like Management you may bring him to the Turnings and strait Turns with little difficulty, and indeed perfect him for any considerable Exercise; in the performance of which, he ought likewise to be considered more than what I have formerly mentioned.

When you intend your Horse for Travel, Sport, or the like, more than what I have already given directions as to Managing, that he may well endure, and answer your expectation, observe to dress him over Night in the same manner as has been directed for the Running-horse; and having by you Dogs-grease or Neats-foot-oil, anoint him therewith especially his Joynts, suppling it in with your warm Hands, or a warm Cloth; and in so comforting the Nerves and Sinews, you will oblige the Horse to hold out much better; and in the Morning give him three quarts of well sifted Oats, sweet and good, with a quart of Beans split, and the husks taken off, and afer them a quart of Ale: and so being accoutered to your mind, ride him forth a Racing-pace, till you find his Joynts very pliable, which will be within a Mile or two riding; then mend his pace, and by degrees put him to either Amble, Trot or Gallop, as best shall please you till you come to your Inn.

Being come to the end of your Stage, if the Horse sweat take off his Cloaths by degrees, and with a piece of broken Sword, Scithe, or edged Lath, scrape him all over, and after that rub him with dry whisps or woolen Cloths; then pass your Hand over him, clean his Pastern and Fetlocks from dirt and gravel, pick his Feet, and cast a Cloth over him, and give him his potion of Meat, which must be a third part more at least than what his usual allowance is when he stands still; and if you find him very hot within, you may give him Water and Ale a like quantity, both being a little warmed, especially in Winter-time, and anoint his Limbs with the same materials, and in the same manner I have before mentioned; and if you find his Breath short, give him an ounce of Hempseed well bruised in a Glass or Canary or warm Ale, ever forebearing to wash your Horse when he is hot, that is, to ride him into a Pond, or the like; but rather if he be exceeding dirty, warm a pale full of Water, and with whisps rub him gently clean; or having first rubbed off the dirt with dry whisps, you may cleanse the rest with a Brush.

And this much I thought fit to lay down as a Supplement to what has been formerly spoken, not much different in the cases of Ordering and Management; and so proceed to Particulars and Generals or another Nature, tho' tending to the same Centre.

Chapter II

How the Jockies make old Horses look young; a lean Horse artificially and naturally, how fatned by Jockies. A Remedy for Restiffness, Neighing, and the Vice of Lying down in the Water, the Art of making Stars, Snips, Blazes, setting on false Ears, Tails, Manes, &c. with a discovery of many other Secrets.

To make a Horse that is really old, seem young.

Rub his Teeth with a Pumice stone, and the Powder of burnt Allom; which rendring them white, take a small Iron, which being crook'd for the purpose, burn in the tops of the two foremost Teeth small holes, so big, that a Wheat-corn may enter on each side of the neither Jaw; and on the Tushes do the like, sining them with a Dodkin till the black Scale come off, and the Teeth in that place look brighter than in another: which done, if the pits above the Eyes be hollow, with a sharp Pen-knife or Lancet slit the skin, being before raised, and hollowing it as much as you can by working of your Fingers, put into the slit a Duck or Crow quill, and blow them up one after another, that the hollowness may fill with wind; which entering into the Cavities of the Skin, will after having been stopped up for a time with a Plaister of *Diaculum* or Bees-wax fix there, till by sweat or extraordinary labour it works out: if the Temples are crooked with a sign of Age, lay to either of them a Poultis of Hemlock and Camomil fried in Linseed-oil; and it will so far contract the Blood to fill them, that for many Days they will appear strait: And then for the Hoof which in case of Age will be seamed or rugged, take a Rasp or File, and having well smoothed it, anoint it well with Oil of Turpentine for a Day or two, and it will look very comely; but in this case the Horse must be disposed of within a Week at the farthest, or else the defects will remain.

To remedy sundry Vices in Horses.

A Horse subject to lie down in the Water, how to remedy it.

In this case you must consider the Horse to be of a hot constitution, begotten or produced under the fiery Signs of *Leo* and *Scorpio*, or else much overflowed with Choler, and therefore ever desirous to cool himself, and thereby rendred in a manner unserviceable: which Vice to remedy, Ride him into a Water up to his Knees and suffer him to lie down; then having three or four lusty Fellows ready with Boots on, let them seize his Head, and hold it under Water whilst another beats and belabours him; and this do till you find the Horse almost stifled; and in repeating it three or four times, the terror of it will work, that the Horse will fly from the Water more than ever he coveted it, and hardly suffer himself to be watered, unless in a Pail or Trough. This likewise may be remedied by Bleeding and Purging your Horse Spring and Fall, whereby the Humours that occasion the extraordinary heat and disorder may be wanting whereon to feed or contract, having the power and force of inflaming his Body, *&c.*

A Tired or Restiff Horse, to remedy.

For the first of these, being subject to Tire without any extraordinary cause, stamp a handful of Nettles, and pour the Juice into either of his Ears; and then take a couple of small pebble-stones, and put in after it, tying or sewing up the Ears, as advantagiously as may be, and with keen Nettles rub his Fundament; laying Hemlock or Ars-smart, called by many Hounds-tongue, under his Saddle, next to his Back.

If a Horse be Restiff, and refuses to go, but will run on one side and go backward, cramp his Stones with a Cord, and bring it up between his fore legs, giving him a twitch when you perceive him forward, and it will oblige him to advance, especially if you keep his Reins even with a steady hand; but if a Mare or Gelding, which yields no such advantage, be troubled with any such Vice, you must have a Crupper with a Brass or Iron-plate fastned a little above the Tail, through which may come two or more sharp points, like Needles, at such a time as the Crupper is moved or strained; and then when you find your Beast practice the Vice, strain the Crupper with your hand, and it will enter the flesh, and force the Beast to leap forward. This Vice generally comes by a Horses not being backed in time, or through the ill management of the Breaker or Rider.

To prevent the troublesomeness of a Horse's Neighing, which may prove disadvantageous to the Master, especially in time of War.

Take a long slip of red Cloth, dip it in the Oil of Linseed, and strew on it the Powder of Elecampane-roots, and on that burnt Allom; fasten it round the Horse's Tongue as near to the Root as may be; and so long as it there remains, you may secure yourself, your Horse can make no noise. And this Art *Darius* the Great *Persian* King, was said to use in the Horses of his Competitors for the Kingdom, by Bribing their Grooms, when the Election depended upon the Neighing of the first Horse.

If a Horse be dull, and will not feel the Spur without much wounding, take the following Directions to make him go very nimble with or without a Spur.

Scrape off the Hair in the Spurring-places on either side, lay a Plaister of Rosin and Bees-wax to soften the skin, then prick it full of holes, so that they may pass through it; which done, take Allom and Copras and rub therein: or you may do it with Powder of Glass beaten very fine, and these entering the holes, will by a little festring, cause such a soreness, that he will run forward upon the last touch with your heel; which having a while used him to, you may heal the Sore with Hogs-grease, Bees-wax and Olive-oil made into an Ointment, with the Powder of Alloes.

To make a lean Horse artificially fat, or to seem so to the Buyer.

Take a Horse lean, but not extraordinary old, rub and comb him well, put him into a warm Stable: Then

Take a pound of Anniseeds, the like of the Powder of Licorish, half a pound of the Flower of Brimstone, and half a pound of Dates stones, with six ounces of the Powder of Elecampane-roots; bruise them well together till they may be reduced to what fineness you think convenient, then with a peck of fine Flower, two quarts of Milk, and the Yolks of a dozen Eggs, make them up into Balls as big as Pullet's Eggs, give him four of these in a morning, and after them a quart of new Wort, then give him half a peck of Oats, and after that a Mash made of Bran, boiled Barley and Lupins; giving him, in all circumstances, the like in the Evening, blowing up his Flanks, and the hollowness of his Eyes with Quills, suffering him to drink but very little, and so in a Week or ten Days he will look very plump and fair to the Eye; but, being afterward neglected, or kept at hard Meat, the spongy kind of Flesh or Fat so gotten, will suddenly fall away, and leave perhaps the unskilful Buyer possessed with an opinion, that the Horse is bewitched.

To make a lean Horse really fat, the best and cheapest way.

In this case, as in many other, People are generally wedded to their Opinions; but Experience being the best Master, I shall deliver therefore what has been faithfully proved and found effectual: and so, to bring your expectation to a period,

Take your Horse from Grass or Soil; and, if the Season admit, Blood and Purge him gently, and so the Crudities, that hinder the kindly operation of the Nutriment may be removed: which done, rub and loosen his Skin, and wash him all over with the Decoction of Hysop, Savin, Rue and Fumitory; which will not only cleanse him from Scurf, and other contracted Filth, but render him a kindly heat and breathing: Then give him each Morning, before his hard Meat, a Ball as big as a Tennis-ball, or such a one as he can conveniently swallow, made of Honey, Rye-flower, the Powder of Licorish, burnt Allom-powder, and the Powder of Harts-horn: And, after his hard Meat, which must be given in due proportion, Water wherein Dandelion-roots, Rosemary and Fennel have been concocted, and in it Bran or Flower dusted; and let his hard Meat be Oats, Splent-beans, and sweet Hay, airing him Morning and Evening, and accommodating him with fresh Litter and good Dressing; by which means, in a Month, he will be fat and lusty, have sound and solid Flesh, and be fitting for any Company.

To make the Hair of a Horse, that stands rough and staring, smooth and sleek.

If yon would have your Horse smooth and handsome, so that your self and others may take delight in him,

Take a French Brush and rub his Hair the contrary way, fetching out by that means the Dirt and Scurf; so that the Horse being freed there from, the Hair may lie closer and smoother; then draw your Brush and Curry comb the right way, and having laid the Hair as smooth as it will be; then, if the Weather be seasonable, let him Blood in the Neck-veins; and after that pass over him with your Hand, or a Cloth dipped in Oil of Bays, and, at convenient times, wash him with the Decoction of Camomil, and keep him well dressed, and at good diet.

To make Hair come where it is thin, or take it away where it is thick.

The Hair being thin, which is uncomely in a Horse, Take the Ashes of Fern four ouncs, the Ointment of Marshmallows two ounces, a Dram of the Oil of *Petrolum*, and an ounce of the Powder of Bithwort-roots: wash or anoint the place with them, mixed with a like quantity of Oil and Wine, adding thereto an ounce of the Honey of Roses; and continue so to do for a Month together: Or, for want of these, you may wash the Horse with a Lye made of the Ashes of Peasestraw, wherein the green Husks of Walnuts and Red Sage have been concocted.

To take off Hair, Take Soot of Wood two ounces, Oil of Tartar two drams, the Calcine of Egg-shells half an ounce, with an ounce of unflaked Lime: make them into a Plaister with Oil of Spike, and apply it to the place you design to have bare or thinner; the Hair at that time being close clipped.

Stars, Blazes, Snips, what they are, and how to make them for Ornament or Disguise in any part of the Horse where it may be conveniently situate.

These are held, by the Curious, to be great Ornaments to Horses; and therefor many have studied how to make them, and left sundry Directions, as their Opinions, how they ought to be made. But since they vary, and many of them upon trial have proved frivolous and ineffectual; thereby I shall only give the approved manner and method of making things of this kind:

First then, if you would have a white Star in a Horse of a different colour, take up the Skin, where you endeavor to situate it, with a Bodkin or fine Lancet, round about, or as you design the Mark; and making a piece of Lead, or divers pieces, the better to get them into the hole, that you within side the hollowed Skin may be framed to the proportion, dipping those pieces in Oil of Spike or Turpentine; by which means the Skin being separated from the Flesh, and receiving no manner of elumentary moisture, will shed the Hair: which when you perceive, take out the Lead, and clap it close to the Flesh, having anointed the Wound with Honey of Roses, and the next Hair that comes will be of a perfect white. Now there are those that take up the Skin by running two Pins or Bodkins cross-ways, and binding them over with a Pack-thread; till the Skin be drawn up like a Purse, that so the Skin may mortifie: but I hold the first the best.

To make a black Star, Blaze, or Snip, in a White Horse.

Take of Souters Ink half an ounce, four scruples of Oliander-wood beaten to Powder, the Juyce of Hemlock half an ounce, Oil of Ebony two drams, and Vitriol two drams: often with these, well incorporated, anoint the designed place. Or, for want of these, Take Galls, the Rust of Iron, Oil of Tartar, and the Ashes of Fern-roots, of each a like quantity: bruise them well together, and apply them Plaister-wise till the Hair sheads; and then observe the next that comes will be black.

If you are desirous of a *Red Star*, then take *Aqua-vitæ* and *Aqua-fortis*, of each an equal quantity; adding two penny weight of Quick-silver: incorporate them as well as may be with two drams of Tartar, and as much of the Flower of Brimstone; and by anointing the place designed therewith, the Hair will become Red, but will continue no longer so than till the casting off the Coat, at what time you may in like manner renew it.

As for Blazes and Snips, the one is made generally on the middle of the Forehead descending, and the other on the Muzzle, between two Nostrils, or somewhat higher, and may be made by the means aforesaid, in what manner and form you shall think convenient, to imitate those that are natural ones: and so you may change Saddle-spots. And seeing those that commonly pass for Stars are only round Spots, I have thought fit to give further Directions on this occasion.

If you intend to make an extraordinary Adornment in the Front of your Horse, you may make a Star, as the form in the Margin, by shaping fine, flat, though taper, pieces of Lead, so that the points extending every way under the Skin, when gently raised, the basis, or broad ends may meet in a centre just in the middle of the Fore-head; the Lead being anointed with Oil of Spike or Tartar, and so suffer it to remain till the Hair is fallen away; and then, taking out the Lead, suffer it to close, having well anointed the inside with Deers-suet and Oil of Roses; and so continue to do till you perceive the Hair well closed, and the white Hair will come undoubtedly in the form of the Figure, which you may make lesser or bigger at your direction: and in this form you may make a black Star, according to the Directions for black Stars.

Some are of Opinion, that a black, or any dark-coloured Horse, may, for Ornament-sake, have four white Feet, as they usually are termed, made artificially; and they say it may be done by the Application of the following Cataplasm, or rather Plaister, *viz.*

Take *Resalgar* a dram in Powder, put to it a dram of the Oil of Tartar or Vitriol, with an ounce of the Oil of Harts-horn, and two ounces of the Juyce of the Roots of Lillies, as much of that of Celendine, make up to a thickness, with the Grease of a Hedge-hog or Urchin, applied Poultis or Plaister-wise: by which means, the Hair being taken up by the Roots, and the Flesh for a time mortified, which may be restored by suppling Ointments, the Hair will come again as white as Milk: Nor am I much different in Opinion; but this being a thing rarely required, and seldom experienced, I leave it to be tried by those that desire it, and proceed to other Matters: But by the way I shall say thus much, that if the Feet by the aforesaid Experiment can be changed in Colour, then consequently the Tail, Ears, or any other part of the Body, by the like application, may be the same.

To make a Blaze-Royal.

In this case, by reason of the intricacy, Lead being very difficult to frame, or at least-ways to put into the skin, so that one part may touch, and the other not; for so it must do, or the Mark will be all of a colour, with a fine pair of Scissars cut the Figure in the Margin as near as exact as may be, to the Skin; and having made a Lye of Urine and Soap-boiler's Ashes; those which appear for the black Stroaks in the Figure, and which you must only cut down, anoint; and so do three or four days, which will stay the Hair in the root; then take *Aqua-fortis* half an ounce, Oil of Sulpher the like quantity; add to them two drams of the Powder of Crab-claws calcined, and with it anoint the places with a small Pencil for four or five days successively, and the next hair that comes in a white Horse will be inclining to black, and that in a black horse to white, and so in any coloured Horse it will alter and chance the colour proportionably. But if you would have this plain, it may be done with Lead, as the Star will make appear; and will not only prove an excellent Mark to know a Horse by from others when at Grass, or the like, but infallibly to describe him by when lost, that without much difficulty of charge he may be found.

The loss of Ears, how to supply.

If your Horse has lost his Ears, and thereby becomes uncomely, you may supply them with those of a dead Horse, by sewing and glewing them on artificially, that the Horse by moving the roots of his Ears, shall move them as if they were naturally his own. But as a farther Mystery, some have held, and indeed it bears the face of a Possibility, that the stumps of the Ears being pared, so that being taken off to the Quick, as near the roots as may be, and the Ears of another Horse new cut off, clapped on warm, and sewed down whilst the Blood issues from the stumps, they will by being anointed with Oil of Juniper, or Bays, by a kind of Inoculation graft themselves to the stumps or roots, and become natural. And this they hold likewise may be done by Teeth, in putting a Tooth just drawn into the place from whence a stump is just drawn likewise. This also they hold may be done by the Tail of a Horse; but not having been proved by me, I leave the Experiment to the discretion of the Reader.

False Manes there are likewise to supply such as fall off, and are not altogether improper, because they serve for Ornament as well as the best, and these are taken off from dead Horses skin and roots, and being takked and dried, so that the Leather becomes pliable, and may be shaved so thin on the fleshy side, that it will let close to the live Horse's Crest, off from which the hair is shaved or taken by the roots, with Applications, it being first clapped on with Glew or

Pitch, may be sewed to the Horse's Skin, in such a manner, that the hairs of the one, and the other rising over the seams, it will hardly be discerned, especially at a distance.

And thus have I laid down such things as I thought convenient of this kind, not to deceive the Buyer; but to the end, that he knowing them, may not be deceived. And as a Caveat, bid him farther beware of logging a lame Horse with Lead to make him go as if he was sound; as also the Spring-saddle, which by the Rider's leaning forward pricks the Horse so intolerably upon bearing between the Shoulders, that if he be never so dull, he will caper and jump as if it proceeded from his high mettle, which hidden cause frequently deceives the Unskilful.

Chapter III

How to set a Horse off for Sale to the best advantage, by Trimming, Washing, &c. as also Directions for the Management of a Horse in Hunting, relating to his Leaping, &c. with other things and matters worthy of note.

Your Horse being repleat with Hair on the Mane, Tail, and sometime, especially if he be of the Dutch Breed, on the Legs or Pasterns, so that thereby he appears rough and unseemly to the Eye; I say, if it so happen whereby sometimes the unskilful Buyer without a cause is discouraged, having your scissers, razor, comb, brush and sponge in a readiness, as also your tweezers or curling iron, comb well his Foretop, Mane and Tail, and with your brush and a comb settle his Legs, if occasion require it; then dust him over, and do the like again after that, beginning at the Foretop with your razor, shave away the short bristling Hairs underneath it, that cause it by the means of their stubbornness to brustle and stand staring; as likewise those that grow unsightly about the roots of the Ears: which done, divide the Foretop into two equal parts, and then clip it with your scisssers a slope between, and so turn it waving up in two divisions with your tweezers, and fasten it with a skewer; and so do by his Mane and Tail, clipping off the ends of those Hairs that hang unseemly, long, or out of order, so that in a short time they will appear comely in waves or ringlets; and thus, although the

Mane falls two ways, you may reduce it to one side, &c. As for the Legs, having clipp'd away the superfluous Hairs, smooth them down with Oil of Bays as hot as may be endured; and in doing so two or three times, you will find the Hair lie close.

How to make a Ball, wherewith a Horse being well lathered and smoothed down, shall look exceeding sleek and comely.

Take of Cake-soap a pound, Fulling earth four ounces, the Lye of Wood-ashes a pint, and Almond-flower two ounces; boil them together over a gentle Fire till they become a gellied thickness, then suffering the mass to cool, make it up by drying it in an Oven or Stove, into a Ball, and with it make a Wash or Lather of new Wort, if it may be had, or else with warm Water, and lather him over with it, being before well brushed down, and so suffer him to stand till the frothy part sink of it self; and then with your Hand dipped in Water wherein Gumaraback has been dissolved, pass over his Body, and after that with a fine Woolen-cloth, and then with your Hand again till he be dry; and he will appear wonderful sleek many Days.

How to manage a Horse in Leaping, taking a Hedge, Gate, Stile, or Ditch, &c.

Many have held this a difficult piece of Horsemanship, as indeed it is; for by ill management in this affair either the Horse or Man has been often spoiled, and sometimes both of them together; and such accidents fall out through the carelessness of the Rider, or ill management of the Horse: Wherefore, to remedy it and prevent danger, let the Rider observe, that he not only sit fast himself upon the motion of rise, but that bearing somewhat forward to give the Horse the more sway, he keep the reins even and steady on either hand; but not too hard, lest the Horse rising too high, may happen to over-set, or lost by such a motion the force of the spring of his hinder Legs, and by that means be rendered less able to cant his Body, or draw them nimbly after him, which frequently occasions a strain or slip, especially at a Ditch. And the best way on this occasion to hold your Rod or Whip, is either between his Ears, or somewhere out of sight, lest it cause him to boggle, by inclining more to the one hand than to the other; or for fear of being corrected, force him beyond his reach. As for your Legs, you ought to keep them moderately close to his sides, but not so close as to press his ribs, for that will oblige him to bate of his strength.

At a Gate or Stile ever observe your distance before you suffer your Horse to take it, which ought to be somewhat more than a Man's pace from it, left by the over reaching of the fore-legs the hinder-legs either strike or stick in the bars, which through the sudden motion occasions an extraordinary bruise, if not the Leg to snap short off.

As for a Ditch, rather chuse to leap against a moderate rising Bank, then upon a descent or low place, because in doing the latter the Horse comes down with so great a force, that he will be either apt to stumble on his Nose, if not well supported by a steady hand, or else to strain or slip his shoulder.

As for the Hedge-leaping, the greatest care required is to shun the Stakes, and the like to be done by the sharp ends of Pails, lest the Horse by an Over-reach gore himself; and in this case consider both the Bank and the Ditch, if any there be, lest you force your Horse upon a thing too hard for him, and by that means be both together left in the mire, or in a worser condition. Nor is it convenient in Hunting to ride a Horse in places where these things are required, unless he be managed at leasure before-hand; for being upon full speed, your Horse seeing others who are managed, undertake these matters of difficulty, and thereby prompted to imitate them, will in spite perhaps of your utmost endeavor to the contrary, resolutely attempt to follow them; for indeed no Creature desires more to Emulate his Like than a generous Horse; nor, according to his understanding, is more proud of winning praise and applause: for, as the Poet has it,

> *When to the Barrs the foaming Steeds are led,*
> *They prancing, seem to scorn the Ground they tread;*
> *And when the Signal's given, either flies*
> *Like Lightning, emulous to gain the Prize; Whilst with their thundering*
> *Hoofs the Earth does shake,*
> *As when the struggling Winds a War within it make:*
> *Nor less, when loud the bloody Blast it sounds,*
> *Do they refrain, o're Spears, o're Death and Wounds,*
> *Through Smoak and Flame, and every dire Attack,*

To rush, by the enraged Warriors back'd,
And face the boldest things that they dare act;
Nay, in the noble Strife, each one does press,
Where Danger may his Courage most express,
To render each his Competitor less.
So that the Lawrel, Conquerers put on,
More by their Horses than themselves is won.
So Great Bucephalus *made it appear,*
Who Alexander *on his Back did bear*
About the World, scattering his bloody War.

Nor have good Horses been less famed and esteemed in all Ages. But having thus passed over what I hope in this kind cannot be amiss, but, on the contrary, prove profitable and necessary, I shall proceed to other Experiments, and things that may yet add to what has been said, that no variety may be wanting in so necessary a Work, nor any thing material be omitted.

Chapter IV

What the Stable to keep a good Horse ought to be; and how he ought to be regarded; the Hoofs how to be corrected and mended in Shooing; and upon other occasions.

As for the Stable wherein a good Horse ought to be kept, let it ever be paved with flat hard Stone, so its long continuance, unwholesome Airs; and let it be likewise laid slantwise shelving, or, as it is vulgarly called, down-hill, that so the Stale may run into a Channel, provided to carry it away; having a Covering of Canvass even with the Rack, to prevent the falling of Mortar amongst the Horse's Provender: or some are of the opinion to have it ceiled, if it be boarded over head, to prevent the Dust or Hay-seeds and many other inconveniences: but, above all, keep Swine and Poultry from coming into it, the Scent of the one, and the Dung of the other, being very offensive. The Rack ought to be made of fir Oak, both the Bearings and the Staves, that the Horse may have the less power to bite them, if he be so inclined; and the Managers be likewise made of firm Plank, well seasoned and dried, having Pins or

Wooden-pegs conveniently placed to hang the Bridles, Saddles, Girths, Cloths, Rubbers, Combs, Brushes, and the like Materials on; observing of conveniency will permit, to have a couple of Windows to open, one to the North, and the other to the South; opening the First in Summer, to let the cool Air, as occasion requires; and the last in Sun shiny Days in the Winter, to let in the cherishing Heat, or fresh and gentle Gales, which will contribute to the vivacity of the Horse, and render him more sprightful; and ever give him for his Litter, Wheat or Rye-straw; but of these two, the former is the best; and if you perceive him subject also to eat his Litter, you must Muzzle him after that he has ended his Provender, lest he otherwise do contract an ill habit of Body; ever giving him his Provender in due time, with convenient opportunity to take his rest at seasonable ties: And if it so happen, as in many cases it is necessary, that you shooe him your self; or if done by another, in giving Directions to the best advantage, you may take the following Rules:

If it happen to be the hinder Feet you undertake, then consider the Toe is ever the tenderest part, and must be pared with a gentle Hand, being left of a convenient thickness, that it may be capable of defending the Quick from Gravel, or being hurt by striking against third Stones; suffering the Shooe likewise to be thicker at the Toe than in any other part; which Shooe must be set on even and clever, not inclining to one part more than another, which you must consider after your having pared the Hoof to an evenness, and struck a Nail on each side, not only be narrowly viewing it, but by suffering the Horse to set his Foot on the Ground; and if any unevenness happen that is not over-great, it may be remedied by striking the Hoof with the Hammer, without the trouble of drawing the Nails; both your Shooe and Nails being made of tough Iron, that they may in no wise be subject to break, and so in a Journey deceive your hopes; the holes for the Nails being made mostly towards the Heels, because of the weakness of the Toe.

In case the Foot has a false Quarter, which looks in a manner like a piece put in, then must the Shooe have on that part a Button, or be a little more raised than the other on the inside, a small space from the quarter nearer to the Toe, that thereby the stress may be taken off the quarter so defective, and so the Horse rendred uncappable of limping.

If by reason of the swelling or standing of the Hoof inward the Horse be apt to interfere, then must so much of it as is convenient be taken away with a Rasp; and in paring the outside, suffer the inside to be somewhat of the highest, that it may oblige the Horse to tread a little outward: or if it cannot be so well done in paring the Hoof, that it may have no opportunity to cut or interfere; and in case the Horse be Hoof-bound, anoint his Hoofs with Oil of Turpentine, and stop his Feet with new Cow-dung fried in Olive-oil and Hogs-lard, making him a Shooe in the form of a Half-moon, that by its openness it may be no obstruction to the encrease or stretching of the hoof. And this may suffice for the Hinder-feet in case of any reasonable Hoof; from which I shall proceed to those of the Fore-feet:

In case the Hoofs of the Fore-feet be well placed and sound, even, firm and tough, then any reasonable shooe will serve the turn; yet in this case must it be pared, and kept from running to disorder; and when you have occasion to travel much in Winter, or in any wet season, it will not be amiss, to prevent its fretting and expanding, to set the edges with a Bat, or other piece of Iron suffering the shooe to swell no more on the toe than heel, because, in this case, the heel is the tenderest part, and must, the rather of the two, rise higher than the toe; and if the Horse be not subject to interfere, you may suffer the shooe, for the preservation of the hoof, to stand a straw's breadth beyond it, even every way; nailing it on in such wise, that the points of the nails may stand in an even round, and seem upon their ringing off and clenching to sink somewhat into the hoof; ever observing to keep gravel from under the shooes. Now there are divers sorts of unnatural and offensive hoofs, hapning through neglect and accident, some of them that will never bear a shooe well, and others that cannot be brought to it without much pain and industry; and the principal of these are,

The rugged hoof, the brittle hoof, the narrow heels, the flat hoof, the broad frush, and the crooked hoof; and to these, shooes must be purposely made and fitted, and the paring be shaped accordingly: the Particulars of which being uncertain or at least too tedious for this Disourse, I refer them to the skilful Farrier, whose proper Business is to order and reduce them, so as they may become practicable and useful.

If the hoof be damaged by reason of any pail that lurks therein, not easily to be found, it must by all means be looked after, and taken thence; and the place of its aboard or lodgment may be found by sundry means and ways, but chiefly by the heat of that place more than any other, or by the Horse's shrinking up his foot.

The place being sound, pull off the shooe, and ope the foot gently with a Buttress or Drawing-knife; and if you can come at it, pull it out; if not, apply the plaister I have mentioned for drawing out thorns, stubs or splinters: or, for your more readiness, thus:

Having in the best manner laid open the Wound, Take of Stone-pitch, Tar, Turpentine and Bees-wax, of each an ounce, and half a quarter of a pint of the Juyce of Garlick; make them over a gentle Fire into a Plaister, and apply them as hot as may be endured; and sometimes it so happens, that by long continuance the Nail in working breaks out above the Hoof; and in such a case apply Honey, Burgundy pitch, and the Powder of Burnt-allom, or a Poultis of Mallows, Camomil and Groundsel, fryed in Hogs-lard, and beaten up with the Whites of Eggs and Rye-meal, and applying it as hot as may be endured; and in so often doing it will break the Skin, so that the Cause of the Grievance may be removed; but by the addition of half an ounce of Verdegrease it will bring away the Corruption, and heal the Wound.

Critical Days, and the Observations thereon.

And now, since it may not be amiss to say something of the Critical Days, observed by many as to Health and Sickness, which relate not only to Humane Constitutions, but those of all Creatures, I shall here insert them.

These Days are accounted such wherein are manifested alterations in realtion to Sickness or Health, Life or Death: And as for the Critick-day, it is ever principally to be observed after the beginning of the Sickness, as, the 3, 5, 7, 9, 11, 13, 17, 21, 29; in which Days, so curiously to be observed, no strong Medicines or Purgations ought to be given; nor is it convenient to let blood, unless great necessity require it. And other Days there are in which many learned Men hold, if Man or Beast shall sick, he shall hardly escape, and that not without much difficulty, which are noted to be these, *viz. January* 1st, 7th. *February* 3d, 4th. *March* 1st, 11th, 13th. *April* 8th, 10th. *May* 2d, 7th. *June* 10th, 15th. *July* 11th, 13th. *August* 3d, 21st. *September* 3d, 10th. *November* 3d, 5th. *December* 7th, 10th. And many add, that if any dangerous Disease happen on the 10th, of *August*, 1st of *December*, or 6th of *April*, it will go very near to terminate in Death. Now there are, on the contrary, other Days held to be good Days, that if a Disease happen on them, there is great hopes of recovery; nay, there is great hopes of succeeding in taking a Journey, or any Labour wherein Man or Beast is concerned; and these are held to be the 3d and 13th of *January*; the 5th and 28th of *February*; the 3d, 22d and 30th of *March*; the 5th, 22d and 29th of *April*; the 4th and 15th of *July*; the 12th of *August*; the 1st, 7th, 24th and 28th of *September*; the 4th and 15th of *October*; the 13th and 19th of *November*; the 23d and 25th of *December*. And these indeed by the Ancients, were held in wonderful esteem; they generally taking their measures thereby, as to what I have said in relation to them; nor are they to be neglected or slighted by Practitioners in Physical or Chyurgical Matters, either in relation to Man or Beast; no, nor in their undertaking, relating to labour, or the like, but have

Some further Considerations upon the Cause of Diseases, and how to remove them, Physically discussed, &c.

The cause of a Disease proceeds principally from an effect against Nature, and happens either External or Internal; the External is that which is outwardly visible, and consists generally of what comes by strokes, bruises, wrenches, scalps, shot, or any matter of wound: Those Internal are occult, or hid within the body, and consequently the less discernable, and are divided into an Antecedent and Conjunction: The Conjunction is that which is nearest, and immediately causes the Disease, and is generally held to be the blood which causes the flegm; nor is it reckoned the Antecedent doth actually cause the Disease, but procures matter, and stirs it up almost to the creating of a Disease; but between it and the Disease are some Causes placed, *viz.* abundance of humours and ill digestion; and therefore these things ought to be chiefly considered before any one absolutely attempts to dispel the disease, by reason diseases are first cured by removing the Cause antecedent, and after that the Cause conjunct: And as for the External Diseases they ought to be known, because they breed Diseases Internal, and very much change the blood, and with much diligence are therefore to be sought out, that so the Practitioner may be brought to the perfect knowledge of Internal Diseases; and these External Diseases are not either to be avoided or amended, but necessarily penetrate the

Body, as Air infected, Meat, Drink, Labour, Sleep, Watching or Wakefulness, Repletion and Evacuation, and some disorders that happen, or to be avoided as unnecessary; as Bruises, Strains, Slips, Wounds, Strangling, or the like.

As for the true cause of a Disease, it sometimes proceeds from corrupt matter whereof they are generated: or when either the Sire and Dam is infected with any Disease, the corrupt quality of which coming into the Seminal Vessels, and transmitting the like to the Creature generated, making it hereditary; and at other times it proceeds from bad usage, bad feeding, heats, colds, and many more which I have named in the First Part. And again, as I have here hinted, internal Diseases may proceed from external Strokes, Bruises or the like. And thus much for Physical Observations of this kind; from whence I proceed to give an accurate Relation of the Spirits, by which the Frame of Life is supported; which take as followeth:

The Spirits, what they are, with their Office, &c.

The Spirits is the aerious and subtle Substance of a Body generated of the most pure and thin Blood, and is the original Mover and Supporter of the Members, giving them power to perform their office, and is seated chiefly in the Brain and Heart, from whence it dilates itself by the means of the Nerves and Arteries into all parts of the Body, and is divided into three parts, *viz.* Animal, Vital and Natural; the Animal peculiarly claims the Brain for its seat, for there it is prepared and made, and from thence defuses itself by insensible ways to the Eyes, Ears, and the like.

The Vital Spirit is chiefly seated in the left Ventricle of the Heart, and has its passage through the Arteries, being made of the evaporation or sweet-breathing of the purest Blood, and is furnished with matter to rarifie it from the Air that is drawn in by the Lungs, and by its motion thro' the frame of the Body is the conservation of natural heat.

The Natural Spirit is ingendred in the Liver and Veins, and is wonderfully instrumental in the concoction of the nutriment, and turning it into blood, and then is farther imployed in distributing it through the Veins that branch from the *Vena-cava*, into every part of the Body. And if any of these are wasted or expulsed from its proper seat, unless it speedily return, there is no hopes of life: And so in case of Famishing, *viz.* when the nutriment in the stomach is spent, then Nature preys upon the Blood, and after that upon the Humour; and lastly, upon the Vital Spirits; at what time the Body, though alive, is past recovery.

A further Description of the External Parts, &c.

And now, the better to describe, or at least make the Reader more sensible of the Chyrurgical Part of the Book, I proceed to give a brief Relation of the material Parts of a Horse's Body: As,

If The Members are Bodies, ingendred of the first commixion of the Elements, Humours and Spirits, because they are found to consist of a solid, fleshy and spirituous Substance; and these are distinguished by, or divided into eight principle Parts, as Bones, Sinews, Ligaments, Tendons, Fibres, Membranes, and simple Flesh and Skin; to which may be joined Veins, Fat, Marrow, Arteries, Hair, Hoofs, and the like. Now to come nearer the purpose, a Bone is of substance earthly, dry and hard, the better to support the frame of the Body, and is it self nourished thro' little Pores, by the purest fat, converted through heat into Marrow. A Gristle is likewise termed a soft and pliable Bone, the better to strengthen and move the Members. A Ligament is a tough and more pliable kind of a Gristle, holding or bending the Bones together. A Tendon is the end and tail of the arbitrary Muscles by which the Members are more easily moved. A Fibre is a small Threat, firm and strong, which Nature places in the Muscles to create a right motion, or a motion every way; for as the right Fibres draw forward, the transverse put back, and the oblique hold fast. A Membrane and a Coat differ only in this, *viz.* A Membrane is the name of a Substance, and the Coat the name of an Office; for where a Membrane invests any part, it is called a Coat. As for the Skin, it is generally called the great Membrane, covering the whole Body, and over-casting the frame or structure thereof, and is made spongy or porous, the better to purge out the excremental Moisture by Sweat. As for the Flesh and Fat increased by nourishment, it is raised and produced from and by the purest sort of Blood and Nutriment: And when between all these there is a concordance and agreement, then is there a cheerfulness in the Spirit, and a harmony throughout the frame of Nature.

As for the Materials you ought to furnish your self withall for the Performance of the sundry Cures I have mentioned, if you are unskillful in knowing them, or any of them: As for the Herbs and Roots, a Herbal will direct you; and for the Minerals, Gums, Seeds, Barks, Oils, and the like, you may consult some Druggest; for should I go about to describe them particularly, they would take up a large Volume; and when that was done, you not perhaps much the wiser, unless the Portraictures were likewise displayed: Wherefore let it suffice that I have spoken intelligibly of all things which I conceived necessary or dependant on this Subject; and have, according to my promise, exposed to your view, and left you to your consideration sundry rare Receipts and Experiments, never before in Print; and such as, I doubt not, will be Approved by those that vouchsafe to try them, and turn not only to their Pleasure, but Advantage, and be a means to preserve this generous Creature in the state of Health and Strength, and render him tractable and obedient to his Owner: and in such hopes I remain further to serve you, as opportunity or occasion will admit.

A Treatise,

Shewing the

Diseases and Cures

IN

Cattle

Part III

Seeing these Creatures are very useful and profitable to Mankind, it is altogether necessary to prescribe such Things as are necessary for their Improvement and Preservation: As for the Breeding and Managing there are few ignorant, whose Affairs lead them this way; therefore the Diseases incident to them are here more properly as an Appendix to the foregoing Work, which by this time we hope has given sufficient Satisfaction.

How to know if a Beast be sound or not, as also to know if an Ox or a Cow be sound or whole of Body.

Go to your Beasts in a morning, which are in the House, before they have meat or drink, and behold the tops of their Noses; if there be standing Pearls, like drops of Dew-water, they are then sound of Body; but if they be dry on the top of their Noses, the Beasts are not then in health. Gripe and Pinch and Ox or Cow, with your hand on the Back or Withers behind the Fore-shoulders; if they are sound, they will not shrink; but if they be not sound, they will then shrink with their Backs, and be ready to fall: This hath been often proved.

How to Fat an Ox.

Ye may quickly fat an Ox with Fetches, or Pease boiled with Barley, or Beans husked and bruised: you may also fat an Ox well with Hay, but not to give him as you give unto a Horse; if you give him in Summer of the tender Branches of Trees, it will refresh him; if you give an Ox only Acorns they will make him scabby, except the Acorns are dryed and mixed with Bran and such like. Also an Ox that you intend to make fat for sale, you may labour him, in fair seasons, once or twice a week, in gentle grounds, and work him now and then; a little Exercise will make him have a better stomach to his Meat; let him eat nothing but Barley and Hay, and sometimes a few Herbs or Vine-branches, or other tender Branches that he loveth, so you shall keep him in good order. Also to fat an Ox, you should give him ground Beans, dryed Barley and Elm-leaves; but more especially his going in the Sun doth make him like the better, and wash him twice or thrice a week with warm water: also Coleworts boiled with Bran doth make their Bellies solable, and it nourisheth as much as Barley; likewise Chaff mixed with Ground-beans is sometimes good for them. Your Oxen are less subject to Diseases than your Horses, yet to preserve and keep them in health, the most experienced Grasiers did use to purge them every Quarter three days together, one with Lupon-pease, the other with the Grain of Cyprus, beaten in the like quantity, steeped all one night before in a pint and a half of Water, and so given; others give them things according to the use and custom of their Country. Again, if an Ox do wax weak and feeble in labour, they do use to give them once a month of Fetches beaten and steeped in Water mixed with beaten Bran: and to keep an Ox from being weary, they do use now and then to rub his Horns with Turpentine, mixt with Olive-oil; but let them take heed that they touch no other part of his Head, but his Horns, for if they do, it will hurt his Sight: also there will sometimes a rising come over the Heart of an Ox, shewing thereby as though he would Vomit; to help the same, you should rub and chafe his Muzzle and Mouth with bruised Garlick, or else with bruised Leeks, and then force him to swallow them down.

An Order or Course how to fat Oxen in the Stall.

Whensoever you intend to fat Oxen in the Stall, being in Summer or Winter, to set them up: if you take them from Grass in Summer, they will hardly fall to eating of Hay of a good while after; but when that you shall take them up, keep them without Meat and Water one Day and a Night, so that they may be extreamly hungry as to forget the Grass; at the first give them a little Hay at once, that they may eat it up clean, and thereby grow still more hungry. On a dry Day take them up into a Stall; for if you stall them wet, they will have (as some Grasiers say) Warnel-worms on their Backs, which will make them lousie, therefore you must use to comb them with a Wool-card or Horse-comb, for that makes them lightsomer and lustier. But some write that 'tis good for the laboring Ox so to be used, but not for the fatning Ox, nor should you let them go forth of the Stall at any time, not so much as to drink; for them they will desire the more to be abroad; and the licking of one another will hinder (as they say) their fatning. But you shall provide that they may have the Water brought to them in Cowls, or else come through the Stalls, as some do use, which is set in a wooden Trough along thro' their Stalls, and with a Pipe of Lead, and a Cock at the end thereof, coming from a Conduit or Cystern, the said Trough is filled twice a day with fresh Water, morning and evening, and at every time before to cleanse the Trough of all the old Water, and so to give them fresh, for Oxen and Kine are Beast that love to feed sweet and cleanly; also you should lay your Trough somewhat aslope, so that the Water may run all forth at the end thereof, in taking forth a Pin to let out the said Water, and then wash the Trough clean, and so give them fresh: thus you may use them morning and evening as long as you fat them. And first in the morning you shall take away all the old Hay, and so cleanse their Water-trough, and give them fresh Water, and then a little fresh Hay again, and so at noon, and likewise at night: and thus use them to feed from time to time. Also it will be best to place and set their Troughs on the further side of their Crib, nigh unto the wall, and to set it two foot high and more, and their Racks likewise should stand of a good height, as of four foot or more, to be made almost as broad beneath as above, for fear of tangling their Horns therein, and the Rack-stands set not passing four inches assunder, yet some do use to feed them on the ground with a Rack, but that we imagine to be more dusty and wastful of Hay. They do also give them, sometimes for change, Wheat and Barley Chaff, with the Gurgine thereof; for that after it (they say) they will drink well. But the Hay is the chiefest Fodder, that will make the hard flesh. Likewise their Standards or Posts to fatten them by, should be made round and smooth, of the bigness of sixteen inches about, and seven foot long, and set them four foot wide one Post from another; you must see likewise that they are set fast and strong, both above and beneath his Neck and Standard, so that the one side of the Neck shall be always close unto the said Standard or Post, by which order of tying to, they shall not at any time bow their Heads so as to lick themselves. And also if you tye them otherwise as Plough-oxen are, with a Sole and a Wythe, which is made like a Yoak, it will be too long a tye from the Standard, and they will both them lick themselves and strike each other with their Horns; therefore the other way of handling and tying is best: some make a light Cradle of Wood, and put it about their Necks, which will keep their Heads from turning back to lick themselves in any part of their Bodies, but it is very uneasie to lie down with: others do also smear them with their own Dung, then cast Ashes, which keeps them from licking. And also for the cleansing and forming of them, they use morning and evening to shovel down their Dung, and to see from time to time that they are clean kept, for that is a very great furtherance to their fatting and liking. And as for the Littering of them, there are some against it, and do not Litter at all, but let them lie on fair dry Planks, and in their own Dung. Yet some are of this Opinion, that to Litter them somewhat under the fore part of their Bodies might do them good. The Keeper must from time to time look to them, and mark if they do eat and drink as they should do; for sometimes there will grow Diseases amongst them in their Mouths, as Barns, and such like, which will hinder their feeding, and kill them in time if they have not help, and are not very well looked to. Some do make holes behind them, and set therein Earthen-pots, even to the Ground, to keep their Piss in, and so cover them with small Boards or Planks, with which Piss they do use to wash the Bodies of those Apple-trees which are inclining for to be Worm-eaten, or Canker-eaten. They also use to cast the said Piss amongst the Roots in their Gardens, for that will kill and cause the Worms to void, and save the Roots from being rotten. Thus much for the ordering and fatting of Oxen in the Stall.

Part III: Diseases and Cures in Cattle 155

How a Man may be rightly informed for to Buy and Sell Oxen.

You are to understand, that Oxen are according to the Region and Country where they are bred; for as there is a diversity of Grounds and Countries, so likewise there are diversities of Bodies, of Courage, of the Hair or Horns of them; for those Oxen in *Asia* are of one sort, those in *Europe* of another: neither are there so many diversities of Provinces, but as many diversities of Beasts; as in *Italy*, in *Capua*, they have White-oxen, of a small Body yet very good to labour at the Plough, and to till the Ground; also in the Dutchey of *Urben* there are great Oxen both white and red, mighty in Body, and of a great Courage: in *Tuscany* and about *Rome*, the Oxen are well set, thick and fit to Labour: likewise in the *Alps* and Hills of *Burgundy*, they are very strong, and can endure all Labour. But nevertheless although they do thus differ, yet the Buyer shall mark and understand herein certain general Rules of Oxen, the which *Mago* of *Carthage* hath given us, and which have been confirmed by the most experienced Buyers of Cattel: He that will buy Oxen, must buy young Oxen well quartered, having large and big members, with long horns, somewat black, strong and big; his fore-head broad, and his brows wrinkled, his ears rough within, and hairy like Velvet, his eyes great and large, his muzzle black, his nostrils crooked within, and very open and wide, the chine of his neck long, thick and fleshy, the dewlap or skin that turneth under his throat, to be great, and hanging almost to his knees, his breast round and big, and his shoulders large and deep. His belly big, compass'd in fatting deep, his ribs wide and open, his reins large, his back straight and flat, with a little bending toward the rump; his thighs round, his legs straight and well trust, his hoofs and claws on his feet to be large and broad under foot, his tail long and well-haired; to be brief, his body to be thick and short, his colour to be red or black, black is accounted the best; also to be gentle and easie to handle or touch, and tame enough to lead. If the young Husbandman would buy lean Kine, or Oxen to feed, he must first see that they are young, for the younger they are the sooner they will feed; let him look well that their hair stair not, but that they do use to lick themselves; and see also that they are whole mouthed, in wanting none of their teeth, for although he have got the gout, and broken both of tail and pizle, yet will he feed; you shall chuse him with a broad rib and thick hide, not loose skinned, nor yet to stick too hard to the rib and sides, for then they will not feed so well. If you buy Oxen for the Plough, see that they are young and not gouty, nor yet broken haired of tail or pizle. Again, if you buy Kine for the pale, ye must also see that they are young, and have such properties as before are mentioned gentle to milk, and likewise to nourish up their Calves. But if the young Grasier be resolved to buy fat or lean Kine, let him first handle thm, and see if they be soft on the crop behind the shoulder, and also upon the hinder most rib, and upon the huckle-bone, and on the notch by his tail; and to see likewise if your Ox have a great cod; and a Cow to have a great navel; for that is a sign that he should be well tallow'd. You must take heed where you buy lean Cattle or fat, and of whom, to know where they were bred; for if you buy from a better Ground than you have your self, these Cattle will not so well like with you: you shall also look that there be no manner of Sickness amongst those Cattle in the Quarter or Parish wherein you intend to buy them, for if there be any Murren or other infectious Disease, it is a great hazard to buy any Beast that comes from thence, for one Beast will soon take a Disease from another.

How to keep a Cow that is great-bellied with Calf.

You must preserve your Kine with Calf, as nigh as you can, from all mischances and dangers from the tenth Month; for when the Cow waxeth great-bellied, and also in Winter, if she be then with Calf, you must nourish her in the Stall, from the eighth Month, because of extream cold, and then for to give her good Meat; but in Summer ye shall give her the branches or tender boughs of Trees. Let her feed in the morning in the Fields, and so milk her, and so give her also in the evening fresh Forrage, when she cometh to the House; and likewise in the mornings before she goes to the Field; and when she hath calved, you shall keep back the Calf when she goeth to the Field: thus use her still as you shall see cause. They do use to labour their Barren Kine after nine Years, therefore they are put into the use of drawing the Yoak as Oxen are.

Of a Cow that wants Milk, having but lately Calved.

To cause her Milk to increase, you shall boil Anniseeds in good Ale, or Wine, then strain it and give it her milk-warm.

Another, Take a handful of the Leaves of the Hedge-vine, called Briony; boil it in Ale or Wine, and strain it, and so give it her.

Part III: Diseases and Cures in Cattle 157

A good way to Cut or Geld a Calf.

You shall cause one to hold down his fore-part, then bind his Hinder-feet with some cord, half a yard assunder, also let his Fore-feet be bound, and let the said Holder set both his knees on the Calf, nigh to his legs, and so cut him gently, and anoint his flanks with some fresh grease, then rub his reins with some cold water mixed with salt, and he shall do well. Some do use to geld when the Calves are young, and some do use to let them run a year or more before they geld them, which is counted more dangerous. After they are gelt keep them in good pastures, that they may be the readier and stronger to labour at three years. Also if the Calves be not gelded within one year, they will prove great. If there grow any Imposthume after the gelding, burn his stones to ashes, and cast the powder thereon, and it will help him. Some are more Astrologically given to observe Seasons and Planets, and think it best to geld them in *Autumn*, when the Moon is on the decrease, and the Sign from the place: In Spawning, Gelding, Cutting, or Letting blood, these signs are most dangerous, if the Moon have power over them, as, *Taurus, Leo, Gemini, Virgo,* and the latter part of *Libra,* and *Scorpio*; also the two Signs governed under *Saturn,* as, *Capricorn* and *Aquarius*; the rest are all good, as, *Aries, Capricorn, Sagitarius,* and *Pisces*; be sure also that the Moon be not in them.

To Rear and Breed Calves for Increase.

If you will breed Calves to make young Bulls, take no Calf that is calved within the Prime, which is counted five days after the Change; for these, as some Husbandmen report, will not prove well. Nor likewise any Calf (or other) for then Calves are not good to keep, but to eat and sell: and among a hundred Calves two shall be sufficient for to make Bulls; as for the rest, it will be best to geld them, after that they are calved. It will be necessary for Husbandmen to rear as many Calves as they can conveniently keep to maintain their stocks, and chiefly those Calves that do fall betwixt *Candlemas* and *May*, for in that season their milk may be best spared, and by that time there will be sufficient grass to wean them, and by the Winter following they will be strong enough to save themselves harmless amongst other Cattel, having now and then small helps: And also the Dams by *June* shall be the readier to take the Bull, and to bring other Calves in the time aforesaid; and if a Cow tarry till after *May*, e're that she calve, the Calf will be too weak in the Winter following, and the Dam will not be so ready to take Bull again, but thereby oft times grow barren. Also to rear a Calf after *Michelmas*, and to keep the Dam at her meat, as they do in some Countries, would be costly in the Winter-time; as a Cow abroad will give more milk with a little grass; than with fodder lying in the close house, or fed with hay or straw, remaining in the stall; for the dry and hard meat doth diminish more milk a great deal than grass. As for those Husbandmen that have but small pastures, or none at all, they must do as they may, though in my mind it were better for them to sell their Calves than to rear them, whereby they may save the milk for more profit to the keeping of their houses, and the Cow will rather go to the Bull again. Also if the Husbandman do go with an Ox plough, it will be convenient for him to raise two, or rear two or three Cow-calves, to hold up his stock, if he can so do, and it will be the more profit; as also it is better to wean Calves at grass than at hard meat, if they were at grass before: And those that can have several pastures for their Kine and Calves, shall do well, and rear with less cost than others. The weaning of Calves with Hay and Water will make them have great bellies, because they do not stir so well therewith as with Grass, and they will the rather rot when they come to grass: and again in Winter they should be put into houses rather than to remain abroad.

To help the Garget in the Throat of a Beast.

If the Garget be in the Throat of a Beast, it will grievously afflict him; this Distemper is commonly taken through some great drought for want of Water; it will cause a swelling under the shoulders and sides thereof. The remedy is, you shall cast him, then cut and flea the skin on both sides as far as any swelling doth appear; so done, take of the whitest sifted Ashes that you can, and mix them with the Grounds of Stale or old Piss, and stir them both well together, then wash the fleshy sore therewith.

To Cure the Garget on the Tongue.

The Garget of the Tongue of the Beast, Ox, or Cow, is a certain swelling under the root thereof, which causes the Head and Face to swell, and to froth also at the Mouth; the Beast will then forsake his meat, often gulping in his throat. The Remedy is, you must cast him on some straw, and then take forth his tongue, and with the point of a stump knife slit along the middle vein under an inch right from the root of the tongnue, and there will come forth black blood and water, which proceeds from the gall; then rub the place with Salt and Vinegar, and he will recover and do well. This hath often been proved.

Against the Garget coming by any Push.

Whereas the Garget is in some Cattel, from some bruise, or some push, you shall cut a hole where the bruises are, and make it hollow to the bottom thereof: Some do cut and race the skin as far as the bruise goeth, and then take, and have ready of beaten Garlick and the tops of the sharp Nettles, with some Rusty-bacon on the outside, beat all well together, and then put it into the hole: then you must bathe it twice a day, as followeth, Take the Grounds of Ale or Beer, and the Soot of a Chimney, of white sifted Ashes, of Black-soap, if you can, mix all these together, stir it well over the Fire, and make it warm; then bathe and wash the sore place; use this morning and evening till it be thoroughly whole. This is an approved Medicine.

Against the Garget in the Maw.

The Garget in the Maw of Cattel is a dangerous Distemper, which is got when the Beasts covet to eat of Grabs or Acorns lying under Trees, which Fruit for the most part they swallow whole without breaking or chewing, so that it lieth whole in the Maw, and will not digest; but in continuance of time they will grow and sprout in their Maws (as some say) till they are like to die thereof. The Remedy is, you should take a good quantity of whole Mustard-seed, and mix it with Wine and strong Ale, and give it to the Beast.

Another, Chop and bruise small a good handful of Camomil, and then mix it with Wine, and give it him.

To kill Lice and Ticks in Cattel.

If your Cattel, Oxen or Kine be Lowsie, which proceeds sometimes from some Sickness, or Surfeit in taking cold after a great wet or rain; and sometimes by being too lean, whereby so long as they are vexed with Lice, they will not prosper. The Remedies are, you shall take the Decoction of Wild-olives, mixt with Salt, then rub and chafe the Beast all over therewith.

Another, Take of Quick-silver killed in Olive-oil, and well mixt together, and therewith anoint: or boil it with good Vinegar, and so wash him therewith. And some do sift Ashes on their Backs and the Rain kills them.

Another, Take the Bear-foot Herb, stamp it, and then strain it with Vinegar mixt with it, and so apply it.

For the Murrain in Cattle.

The *Murrain*, much incident to this sort of Cattle, is known by drivelling, running at the Nose, and Mouth, dullness and sinking of the Eyes, pining away, *&c.*

To Cure this, take Fennel-seed, the roots of Angelico, and Sea-thistle; stamp and infuse them over a gentle Fire with Red-wine and Ale; give the liquid part hot, and keep the Beast warm, and two or three Hours after make him a Mash of Wheat, boiled in Small-beer.

For the Flux or Lasks.

Take dried Sloes, bruise them to Powder, add Raisins and dried Grapes; boil them in Vinegar, and give them hot when the Beast is fasting, and keep him so an Hour or two after.

For the Lasks, or Ray in Calves, or Cough in young Bullocks or Heifers.
For the Calves,

Take new Milk, put in so much Renit as may curdle it; warm it over a Fire, and give it them hot two or three times. As for the other, Take a pint of Barley, the yolk of an Egg, a handful of Raisins; boil them in a quart of *Aqua-vitæ*, and give it them as hot as possibly may be endured.

For Scalds or Manginess.

Rub them well with a Hair-cloth dipped in the Juyce of Garlick and Rue, and a Day after with the Concoction of Penny-royal, and Flower of Brimstone, made with Water and Bay-salt.

For any Distemper in the Lungs,

Take Cloves, Anniseeds, Long-pepper, Turmerick and Fenugreek, of each an ounce; boil them in Small-ale, and give half a pint hot in a Morning for a Week.

For Pissing Blood.

Let them stand twenty four Hours without drinking, and then let them blood in the Tail; after that boil Nettles, and Ash-leaves, or the Bark of an Ash-tree, in Spring-water and give it them hot.

For the Taint and Gargets.

Take Urine and Bay-salt, with the Roots of Red-docks; boil them and bathe the afflicted Place very hot morning and evening, and afterwards anoint it with Sheep's Suet as hot as may be endured.

For stoppage of Urine.

Take a pint of French-wine, the Whites of nine Eggs, six Cloves of Garlick; give the liquid part well pressed and strained fasting.

For any poisonous Infection, or pain in the Belly, swelling or internal Bruise.

Take four ounces of the Rhind of Elder, a handful of Longwort, an ounce of Licorish, a handful of Rue; boil them in three pints of Ale, and whilst they are boiling put in a handful of Bay salt, six Cloves of Garlick, and four Eggs with the Shells; add a pint of Beer, an ounce of *Vernice*-treacle, and the like quantity of Bay-berries; give the liquor hot.

For the Sperenges and Staggers.

Let him blood in the Fore-head by slitting the skin across, and after a short bleeding put warm Vinegar into his Nose, his Head being held upward.

For being Hide-bound, which hinders the Growth of Cattle.

Boil Bay-leaves in Water, and rub them over as hot as may be, and then let them blood in the Neck; and after that rub them with Lees of Wine, and Neats-foot-oil.

For the Feaver in Cattle.

Let them blood in the Tail, give them Colewort-leaves to eat; boil Colts-foot, the leaves or branches of the Vine, and the Roots of Violets in Water, giving them the liquid part as hot as may be.

To Cure Halting.

Slit the Claws till they bleed, and bind a Cloth, dipped in Vinegar and Salt, to the place, and keep the Beast in a dry place; and so by twice or thrice renewing it, that disorder will cease.

To Cure the Swelling in Cattle by breaking into fresh Pasture, and over-feeding, or by licking up some venomous Matter,

Take half a pint of Sallad-oil, as much warm Milk, an ounce of Licorish-powder, and an ounce of Mithridate; give them very hot, and keep the Beast in motion, till he voids at both ends, or the Swelling abates.

For the Blain, or outward Sorrance,

Take half an ounce of Turpentine, the like of Verdegrease, an ounce of Bees-wax, and the like of Mutton-suet; make them into an Ointment over a gentle Fire, and apply them Plaister-wise two or three times to the place grieved.

For the Pains in the Bowels.

Take a handful of Fumitory, as much May-weed, or Camomil; boil them in Milk and Small-beer, and give it hot.

For the Quinzy.

Take half a pint of Vinegar, as much Sallad-oil, an ounce of Mithridate; give them very hot.

Of Rams, Ewes, and Lambs: Their Ordering, Breeding, the Cure of Diseases and Grievances incident to them.

To chuse good Breeders for the increase of Sheep.

Let your Ram be of a large stature, of Body long and full bellied, his Tail long and bushy, his Body thick of Wool, his Eyes black and sparkling, his Body broad, his Cods large, though not over-hanging his Loins, and Ears large, covered with much Wool, his Horns large and bending with the tips from him; also a white and clear Palate and Tongue.

The Ewe with a deep Belly, white Wool, soft and shining on her Neck, smooth Horns, large Dugs, her Eyes Gold-colour, her Dugs long and lean, her Tail full of Wool, and a long Visage.

The best time of Covering.

For the best breed, let not the Ewe take the Ram before she be full two Years old, lest the Lambs prove weak and sickly, of a stunted breed and growth; and in this case she will bear from two to seven Years old, but those Lambs, of her two first Years are the best for breed.

How, after Casting, to order your Lambs.

As soon as may be set him on his Legs, and direct him to his Dam's Tear, having milked the first Milk out, because that will prove hurtful to him, by reason of its curdling; if he refuse to take the Teat, open his Mouth and spurt some Milk into it, which will soon make him more familiar: Or, if the Ewe's Dugs be tender, suckle him through a little Horn, which you must provide for that purpose. And, if they grow wanton, sever them with Hurdles, and tie them with soft Bands to Stakes in the Ground, that they may not frisk, spend their Flesh, or hurt each other: Suckle them Morning and Evening, before the Dams go to Pasture: Give them a little Milk and Bran mixed with Flower.

When they are weaned, which, in warm Weather, may be done at seven Weeks end, keep them well fed, lest their pining after their Dams make them pine or throw them into Diseases; and geld them not till three Months old, keeping them warm and giving them Bran in their Water, pretty hot, for five or six Days after. And so, by well looking to and ordering them, fear not to have extraordinary good Breed.

If you design to keep Ram-lambs for breed, take the best and liveliest of the two, where the Ewe has two Ram-lambs at one time; and see that the feeding and bringing up be carefully observed, ever now and then sprinkling some Salt and Fennel-seeds swelled, bruised in their Watering tubs or Troughs.

And thus, having given you an Introduction as to the Breeding them, I shall proceed to the Diseases incident to them and their Cures.

For the Scab or Mange.

Whether this appear without or within the Skin, delay it not, but take a Quart of Man's Urine, boil the Leaves or Bark of Elder and Hemlock in it; then strain it, and add a Pint of Water wherein Tobacco-stalks have been soaked; clip off the Wood pretty close, and wash the place Morning and Evening, as hot as may be endured: Give them Bay-salt in their Water, and keep them from wet Pastures or much green Feeding.

For bruised Joynts, broken Claws, &c.

For the first, anoint them with the Oil of Spike; then bind up the bruised or broken Joynt with a Poultis made of Mallows, Groundsel and Bettony, beaten with Hogs-lard, and fried in it, applying it as hot as can be.

For the latter: If the Claws be over-grown, lame, or broken, pair them, mix a little Tar-water with Hogs-lard, Bees-wax and Turpentine, and with a Plaister of it bind up the bruised Claw that it may be kept from wet and dirt.

For the Rot or Plague.

For this when you first perceive it among your Sheep, having separated the Infected from the Sound, Take the Herb Melliot, Comfry, Rue, Pollipodium of the Oak, and Walnut-tree-leaves or Bark, of each a handful, boil them in a quart of Water and a pint of *Aqua-vitæ* with two ounces of *Venice*-Treacle or *Mithridate*, and give a quarter of a pint at a time to one Sheep, very hot, for four or five mornings, and it will cure them.

For any Disease in the Lungs of Sheep.

Take Lung-wort, Sage and Colts-foot, a Herb so called, bruise them with a root of Galick, and two ounces of Honey, then boil them in a quart of White wine, with a few slices of Licorish; and give a quarter of a pint hot. This also cures Pursiveness, Cough, or Wheesing in Breathing.

For the Head-ach, or pains in the Head of Sheep.

Take six grains of *Assa foetida*, two spoonfuls of the Juyce of Sage, a quartern of Wine-vinegar, and give them the Sheep as warm as may be.

In case of Rheums, Catarrhs, or Coughs.

Take an ounce of Dill-seed, and as much of Bay-berries, a handful of Vervine, and two ounces of brown Sugar-candy; bruise and boil them in a pint of Cyder, or Verjuyce, and give it fasting as hot as can be endured.

For the Plague and Rot, another.

Wash them with Water wherein Balm has been boiled, give them to drink a Water wherein Rye, Sow-thistle and Vervine hath been boiled, and season it with Allom.

For Boils, Aposthumes, or Ulcers, that are not come to a head.

Mix Rye-flower, Yolks of Eggs, and Tar, applying them Plaister-wise, and when they are drawn to a head, launce or prick them, and apply Honey, Allom and Rosin, made into a Salve over a gentle Fire.

For Scabs, or Breaking out of that nature,

Take a quarter of a pint of the Juyce of Hysop, the like of Camomil, and a quart of Water wherein Tobacco stalks hath been boiled, two ounces of Brimstone-flower, a handful of Fern-roots, and a quart of Urine; wash the Sheep with it hot twice a day.

For Pursiveness,

Take a quart of Vinegar, an ounce of Licorish-powder, *Venice*-treacle, half an ounce, *Carduus* an ounce; boil them and give them hot. This does likewise for short Wind, and cures the Swelling in the Belly.

For St. Anthony's *Fire.*

Bathe them with Goat's Milk, and the Juyce of Briony very hot, and give them Water wherein Sage and Endive have been boiled.

To supple broken Joytns, Strains, Wrenches and fractured Bones.

Put those that are dislocated in their right places, then take an Ointment, Bees-wax, Turpentine, Deers-suet, the Juyce of Mugwort, Stone-pitch, and Melliot, softned with the Oil of Earth worms, and bind up the place, suppling it with Ointments as occasion requires.

For Lameness, which is occasioned by too much Wool growing in the fleshy part of their Feet.

Rub between their Claws Allom, Vinegar, and Bay-salt.

Of Swine; their Breeding and Ordering, with Directions for Curing the Diseases incident to them.

How to have a good Breed of Swine.

Let the Choice of your Boar be such a one whose mouth is drawn upward and long, his breast thick and broad, as also his shoulders, his thighs thick and short, white of colour, mixed with sandy spots, he being almost as thick as long, well bristled, and a pair of even cods.

The Sow you make choice of must be long of Body, large and well-bellied, many tears, her buttocks long, her rubs broad, her head little, her snout long, legs short, and white of complexion.

Then the Boar at a year and the Sow at two years old may be put together, when the Moon increases, keeping her after she is three times served, in a warm yard or stye, according to the season, and if in the Winter-time, give her grains, bran, pease and beans, and sometimes roots and cabage, or colwort-leaves. The most convenient time for her taking Boar is in *February*, and then the Farrow will come in warm weather, for those that come in Winter are either stunted or troubled with many Diseases more than the other: Geld them at three months old, if you would have your Pork or Bacon extraordinarily sweet; but if large and fat, let them run with their stones till six months; and after gelding keep them up warm for a week, and give the pollard and ground beans. As for your Sow pigs you intend to fat up, splay them not till nine months at soonest, and be curious in doing lest you destroy them, for if any stitch in sewing up the rim of the belly take hold on the guts, the Sow will pine away, and never grow fat, if she die not soon by it. And now for the Diseases incident to them.

To cure Pains in the Head or Teeth.

First let the Swine blood under the Tongue, then boil Rue, Savin, and Crople-stone in their water, and scatter some sweet Malt in it, and so they will the better be delighted to take it.

For the Head-ach, or Sleepy-evil.

Blood them under the Tongue, and rub the Wound with Bay-salt, giving them for a time Cabage or Colewort-leaves, Teats, Pease and Whey, to drink.

An Approved Remedy for the Measles.

This Distemper is occasioned by surfeiting through unwholsom Feeding, and is discerned by the coming of knots or pimples under the tongue: The remedy is, wash the Swine in Brine or Salt-water, pretty warm, bruise Garlick, Peels or Lemons and Grapes, steep them in strong Vinegar, and give him to drink.

Another.

Rub them over with Water and Salt, mix Parsley-roots with their Meat, and put Allom into their Water, in which steep Rue.

If Agues or Feaver afflict them, let them blood in the Neck-vein, and Tail, and give them warm Water and Bran thrice a day; also Water between whiles wherein Parsnips and Pepper have been boiled.

For the Swine-Pox,

Take an ounce of Mithridate, and as much Pepper, half a pint of Olive-oil, a pint of new Ale, and two ounces of Honey, give them very warm.

Another.

Take, to remedy this, three ounces of Honey, one of Mithridate, a quarter of a pint of Olive-oil, and half an ounce of bruised Allom, in a pint of Lambs or Sheeps Blood.

For Rheums, or Catarrhs,

Take half an ounce of Brimstone, as much of Burgundlapitch, hold his head by force over them, whilst burning on a Chafing-dish of Coles; after which give a drench of Garlick, Pepper and Rue boiled in new Small-beer.

For the Plague, or any Disease in the Melt,

Take three ounces of Honey, two ounces of Bees-wax, an ounce of Ginger, and two ounces of Coriander-seeds, boil them in three pints of Milk, strain the liquid part and give it hot.

To Cure the Flux,

Take Nut-gall two ounces, as much Starch, and a handful of Bettony, half an ounce of Turpentine; boil them in a pint of Milk and a quart of Vinegar, and give it hot three mornings.

For the Belly-ach,

Take an ounce of Long-pepper, a handful of Fennel-seed, an ounce of Fenugreek-roots, and two ounces of Honey, boil them in a pint of White-wine, and a quart of Stale-beer, and give it fasting.

For the Diseases in the Eyes of Swine,

Take Rue, Pimpernel and Vervine, each a little handful; dry them in an Oven, so that they may be powdered, and blow the Powder, mixed with the Powder of Bole-armoniack, into the Eyes grieved, and it takes away Spots, stays Fluxes of Rheums and Redness.

The Vermin-killer:

BEING

A very necessary Family-Book, containing exact Rules and Directions for the artificial Destroying of Vermin.

To Kill Rats and Mice.

Take Hellibore-leaves, mix it with Wheat-flower; make it into a stiff Paste with live Honey, and lay it into Holes where Rats and Mice come, and when they have eat of it, it's present Death. Approved, *Paxamus*.

To make Rats and Mice blind.

Take Tithimalum beaten to Powder, sift it through a fine Sieve, then mix it with a like quantity of Wheat-flower, then put to it a sufficient quantity of *Metheglen* to make it into a stiff Paste, and lay it in the usual places where the Mice and Rats come, and in a short time after they have eat it, you will see the effects, for they will become Beetle-blind. Approved, *Anatiolus*.

To drive away Rats and Mice from a House or other place.

Take the Herb wild Marjorum, called in Lattin *Origanum*, and burn it in all the Rooms of the House, or place where you would be rid of the Mice, and they will immediately depart, and come no more so long as the Scent lasteth. Approved, *Paxamus*.

To gather together all the Rats and Mice into one place in a House or Barn, and kill them.

Take a Brass or Copper-pot, as big as you can get, and put into it the Dregs of Oil about half full, and then set it in the most convenient place in the House, about the middle, and all the Rats and Mice will make their appearance, as if it were to be an Assembly of an Army of Rats and Mice, and you may then throw about the place Pot-ashes, and it kills them all. Approved, *Aborto.*

To Catch Moles,

If you desire to catch Moles, lay before the Mole-holes a head of Garlick or Onion, and they will immediately forsake their Holes, and may be taken by a Dog. Approved, *Albertus.*

To Kill Moles.

Take white Hellibore bruised very small, mix with it Wheat-flower, the White of Eggs, Milk and Wine, and lay little Cakes of it in the mouth of the Holes, and the Moles will greedily eat of it, and it certainly killeth them. Approved, *Paxamus.*

To Kill Weasles.

Take Sal-armoniack and Wheat-flower, mix it into a Paste with Honey, and throw it in such places where the Weasles usually come; they greedily eat it, and it quickly killeth them. Approved, *Cornel. Agrippa.*

To prevent Weasles from sucking of Eggs.

Take Rue and lay it about the places where the Hens lay, and the Weasles will not come near it.

To drive the Snakes and Adders out of the Garden.

Plant in several places of the Garden Worm-wood, and they will not frequent the Gardens. Approved, *Paxamus.*

To kill Snakes and Adders.

Take a large Reddish and strike the Snake or Adder with it, and one blow will kill them. Approved.

To Kill Pismires.

Take the Root of wild Cucumbers, and set them on fire, where the Pismires are, and the Smoak will kill them. Approved, *Agrippa*.

To preserve Plants from the Pismires.

Take Lupin beaten with the Dregs of Oil, and anoint the bottoms of the Plants therewith. Approved.

To drive away Pismires.

Take an Earthen Dish full of Pismires, and the Earth where they are, and make a good Fire, and lay the Earthen Pot on the Fire, and the Pismires will not remain near that place. Approved.

To kill Bugs.

Take a convenient quanity of fresh Tar, mix it with the Juyce of wild Cucumber, and boil them in Pickle, and sprinkle it in the Room, and it will certainly kill the Fleas. *Approved,* Cornel. Agrippa.

Another.

Take *Malenthiam* steeped in Water three or four Days, then sprinkle the Room with it, and you will immediately find the effects. *Paxamus.*

To Kill Lice.

Take Salt-water and rub the afflicted places with it, or Vinegar or Onian, and mix it with Allum and Alloes, and therewith anoint the place. *Alex.*

Another.

Take Hogs-lard, Quick-silver, and Sage, as much of each as is needful, mix them together to a Salve, and anoint the afflicted places. *From an Italian.*

To kill Crab lice.

Take a roasted Apple, and take the Skin and Core from it, and beat it in a Mortar with as much Quicksilver as will make it into an Ointment, and therewith dress the afflicted places. *From a good Friend.*

For Nits and Lice in the Head.

Take three Ounces of Oil of Olives, one Ounce of Wax, three Drams of *Stavesacre*, and as much Quicksilver, and of these make a Salve, and anoint the Head all over, and it certainly kills the Nits and Lice. *Approved,* Cornel. Agrippa.

To Kill Caterpillars.

Take Fig-leave-ashes, and cast it on the Root, and it destroys the Caterpillars. *Anat.*

Another.

Take Ox-piss and Lees of Oil, and boil them together, and it kills the Caterpillar infallibly, if you cast it upon the Trees or Bushes where they are. *Anat.*

To kill Flies.

Take white Hellebore, and steep it in sweet Milk, mix with it Orpiment, and sprinkle the Room and places where the Flies come, and they will die. *Approved.*

Another.

Take Allum and Origanum, beat them and mix them with Milk and sprinkle them, as before. *Approved.*

To gather the Flies together.

Take a deep Earthen Pot, and lay in it beaten Coriander, and all the Flies in the House shall be gathered together. *Approved,* Anatolius.

To keep Cattle from Injuries by Flies.

Anoint the Breast with Oil wherein Bakelor hath been boiled, and the Flies shall not come near him. *Approved.*

The Art of taking Fish.

Take Time, Savory and Elder-leaves, or each a like quantity, take the Suet of an Ox or Sheep, the Lees of Wine, and make this into a Mortar, then break it into little Balls, and throw them into the Water an Hour before you intend to fish. *Approved,* Albertus.

Another.

Take the Blood of an Ox or Sheep, then take the Dung of a Sheep out of the small Guts, while it is hot, and mix it with the Blood, and make small Baits of it, and let them dry in the Sun, and cast the Balls into the Water, and they will gather all the Fish together.

Approved, Cor. Agrippa.

To take all sorts of Birds, Fowls, &c.

Take such Seeds as the Fowls or Birds are wont to feed on, and lay it a soaking in the Mother of Wine, mixed with the Juice of Cicute, and when it is well soaked, throw it in the places where the Fowl or Birds feed, and they will be presently drunk, and lose their senses, and you may take them with your Hand. *Approved,* Albertus.

FINIS.

Suggested Reading

Crabbe, Barb. *The Comprehensive Guide to Equine Veterinary Medicine*. Sterling, 2007.

Curth, Louise Hill. *The Care of Brute Beasts: A Social and Cultural Study of Veterinary Medicine in Early Modern England (History of Science and Medicine Library)*. Brill, 2009.

Gianoli, Luigi. *Horses and Horsemanship Through the Ages*. Crown Publishers, 1969.

Giffin, James M. *Veterinary Guide to Horse Breeding*. Howell Book House, 2004.

Graham, Elspeth, et al. *The Horse as Cultural Icon: The Real and Symbolic Horse in the Early Modern World*. Vol. 18, Brill, 2011.

Holland, Anne. *Horse Racing in Britain and Ireland*. Shire Publications, 2014.

Huggins, Mike. *Flat Racing and British Society, 1790-1914: A Social and Economic History (Sport in the Global Society)*. Routledge, 2014.

Kevil, Mike. *Starting Colts: Catching / Sacking Out / Driving / First Ride / First 30 Days / Loading, Revised*. Western Horseman, 2010.

Landers, T. A. *Professional Care of the Racehorse*. Eclipse Press, 2006.

Smith Thomas, Heather. *The Horse Conformation Handbook*. Storey Publishing, LLC, 2005.

Walsh, J. H. *The Breaking of the Colt: A Historic Article on Horse Training*. Kolthoff Press, 2001.

Weaver, Sue. *Storey's Guide to Raising Miniature Livestock: Goats, Sheep, Donkeys, Pigs, Horses, Cattle, Llamas*. Storey Publishing, LLC, 2012.

Whyte, James Christie. *History of the British Turf: From the Earliest Period to the Present Day*. Ulan Press, 2012.

www.ingramcontent.com/pod-product-compliance
Lightning Source LLC
Chambersburg PA
CBHW031945070426
42451CB00007BA/128